Copyright © 2013 by Thomas W. Newbery. All rights reserved. No portion of this book may be reproduced in any manner whatsoever without written permission, except in the case of brief quotations embodied in critical articles or reviews.

To my wonderful wife, Valerie, whose patience knows no bounds...... usually.

Table of Contents

List of Photos, Illustrations and Figures

Introduction

Possible Uses

Tools

Design Considerations

The Jigs

Spreadsheets are Cool!

Metric Measurements

Calculations Required

Designing the Hubs

Designing the Struts

Making the Hubs

Making the Struts

Before Assembly

Assembly

Cover Up!

List of Photos, Illustrations 'n Figures:

Photo	Cover - in my front yard.
Photo 1	Wood dome frame
Photo 2	As "Monkey Bars"
Photo 3	Sewing Machine
Photo 4	Dome in Oregon
Photo 5	#2 Stubby and Wing Nut Tool
Photo 6	Pail and coffee can combo
Photo 7	Two cans w/belt loop hook
Photo 8a	Improper hub orientation
Photo 8b	Proper hub orientation
Photo 9	"D" ring tie-down installed
Photo 10	Door w/strut removed
Photo 11	Door secured
Photo 12	Window on door
Photo 13	Window on plain gore
Photo 14	Vent Cover
Photo 15	Gore template
Photo 16	Opening showing side strut
Photo 17	Door flap stowed
Illus 1	Hub design
Illus 2	Base Hubs Laid out
Illus 3	Base Struts Laid out
Illus 4	Base Hub 'n Strut connection
Illus 5	Longs and Shorts alternating on B-Hubs
Illus 6	Short Struts raised and bolted to P-Hub
Illus 7	Longs raised and bolted to H-Hub
Illus 8	Shorts bolted between Pent and H-Hubs
Illus 9	H-Hub with 2 Longs and 1 Short, bolted
Illus 10	H-Hub assembly mounted
Illus 11	Top Pent Outline
Illus 12	Top Pent Completed

Illus 13   Seam Allowance Marking Tool

Illus G1   Gore Template 1
Illus G2   Gore Template 2
Illus G3   Gore Template 3
Illus G4   Gore Template 4

List of Figures:

Fig 1   P-Hub drawing
Fig 2   H-Hub drawing
Fig 3   Base hub drawing
Fig 4   Short strut drawing
Fig 5   Long strut drawing
Fig 6   Strut cutting jig
Fig 7   Strut end squish marking jig
Fig 8   Crushing squish in-same-plane jig
Fig 9   First strut hole drilling jig
Fig 10   Second strut hole drilling jig
Fig 11   Alternate second hole drilling jig
Fig 12   Calculating strut squish angle
Fig 13   Strut squish angle bending jig
Fig 14   Wing nut tool

Introduction

As I remember it, back in the early '80's I was browsing through the County Library book shelves and stumbled upon a book with a collection of articles dealing with geodesic dome structures and their construction. I was hooked. The dome structure was so elegant in its simplicity that I proceeded to build my first dome frame out of wood using the information provided in those articles. It was 16 feet in diameter, 8 feet tall, 50 feet in circumference and enclosed 200 + sq ft. (Photo 1).

I covered it in plastic sheeting and a surplus jet fighter's drag chute to shade its interior. For materials, I used 1 x 2 lumber for the struts along with plywood and 1 x 2 lumber for the hubs. 1/4" bolts, nuts, and washers held the assembly together. Wow! What an enormous enclosure! What possibilities! My family and I used it as a "tent" for camping at some recreational property, and I sold two more of them when I set one up at a couple of swap meets.

This book describes my construction of a metal framed dome. The book is meant to be an encouragement for those who might want to build their own dome to suit their purposes. The reader/builder must at all time use reasonable safety precautions and common sense when using tools and materials and in determining the suitability of the dome for its intended use(s). The reader/builder is solely responsible for results he/she achieves and assumes sole liability for any injuries incurred building, transporting and using his/her dome and or its components.

Possible Uses

Possible temporary or seasonal uses for the structure might be as a:

Greenhouse... start your plants early. Covered in translucent plastic, the solar heat gain is amazing.

Tent.... a large tent. Depending upon its diameter, it can enclose the complete campsite of a couple (including the couple's tent, picnic table, and fireplace!) or act as an auxiliary bedroom, kitchen, living room, or recreational room for a family of many.

Auxiliary Room... for RV-ers or campers it could be an additional living- or dining-room, or the outside kitchen. Got family coming for a visit?... you take the dome and they get your master bedroom. (If you're not that fond of 'em... reverse that!)

Temporary work shelter... set it up over your outside work area and remain protected from the elements and warm or cool, as you wish.

Outside spa/pool cover..... with a translucent lower cover for privacy and a clear upper cover for even more heat gain.... talk about extending the season! Heck, you might be able to bathe in comfort all through the winter!

Hunting/Fishing, Ice Fishing camp shelter, year 'round or semi-permanent... My whole frame and cover weigh only about 134 lbs and fit in a package about a foot in diameter and 5 feet long. That means I could drive, boat, backpack, or fly it to a camp site, and when I leave at the end of the season, I could take off the and stow the cover and leave the frame up. Barring any catastrophe like a tree falling on the frame, it'll be there next season! What the heck.. use it as your temporary shelter while building your log cabin! Then use it for, say, a bath house.

As an **ice fishing shack**, with the sun out and a clear cover, you'll be fishing on the ice in your shorts and t-shirt! It may be so warm in there that you'll have to insulate the ice from the interior temperature with some sort of covering, like used pallets!

Skating party/Sledding party warm-up shelter. I remember as a kid going to local park's pond in the winter time after the pond had frozen over. Back from the edge of the pond was a huge warm-up shack with a roaring wood fire going in the center. We'd clomp, clomp over the wooden walkway into the shack, warm up for 15-20 minutes, and clomp, clomp back out to the ice for another half hour.

Sales kiosk.... There are many art "events" and swap meets staged around the country. The structure's frame allows a lot of hanging length for almost anything of reasonable weight. If you make the cover with an entrance and exit as I did, you have controlled access, weather protection, and a little security for your goods.

Sheep herder's living- or bed-room. Beats just having that small cubicle on wheels! Probably make a good radio antenna or antenna ground plane, too!

"Monkey Bars"... This book can serve as a guide to make a dome frame for kids to play on. (Photo 2) One would naturally scale it down so the frame isn't too high off the ground and choose the struts' dimensions to take the strain of the anticipated loads. The hub and strut junctions can easily be wrapped with a soft material to protect the kids from the metal edges. You'll see similar structures' commercial smaller versions at many daycare centers.

Heat storage/source.... This is a very tempting project for me what with the heating costs one faces in the winter. I envision a rock mass in the dome, taking off the heat at the top and routing it to the house through insulated ducting. The northern-facing hemisphere of the dome would be insulated. At night, a second insulated hemispherical shell could be rotated or drawn over the southern side to retain the heat absorbed during the day. The heat would then be drawn off at night for the home.

At Home 'Prepper shelter.... Here in Utah along the Wasatch Front, we're due for a major earthquake. That's primarily why I built my dome. I doubt most homes, including mine, would survive such a quake and still be habitable. What, you're going to call up the nearest hotel and reserve a room? Good luck with that! Live in your car? Try that with a family and kids! What I felt I needed was a shelter that could be stored in minimal space and away from the house (in case it collapses), to be assembled with minimal tools if needed after the 'quake. Mine is in a weather-proof box, an old, used "truck box" and will be placed in a discreet location elsewhere on my lot away from the house. A "rocket stove" for heating/cooking would make a nice compliment to the domes' interior.

Away-from-Home 'Prepper Shelter.... If even a worse(!) event occurs and you must flee the area, for a minimal cost you could have pre-positioned your dome "kit" out in a location to which you'd retreat, assembling it upon arrival. Let's face it, do you have the money to buy, transport, bury, and equip a doomsday shelter? I sure don't. Besides, I'm not exactly desirous of being trapped below ground while under siege by hostile forces.

Trellis.... The frame, covered in vines, would offer a cool, secluded place for contemplation or ….. ?

Sweat lodge/Sauna.... just make sure the frame and material covering the frame are fireproof and the frame is strong enough to support the covering. Heck, in some southern locales, you might not even need a fire in the daytime!

Mobile place of worship... I get inside a dome and it's pretty close to a religious experience for me. There's just something about the enclosure and its shape.

A form for a ferro-concrete dome... just make sure you put it where you want it, 'cause you're not going to be able to move it easily, if at all, once the cement is poured. I'm thinking of a ferro-concrete structure. Plan it so you can dis-assemble the frame from inside and use it again! Rodent-proof **grain storage,** anyone?

Tornado/hurricane shelter... it's shape would seem to be a natural. In ferro-concrete, a dome, well anchored, just might be the most destruction-resistant and immovable enclosure for survival during such violent acts of nature.

Observatory enclosure.... for the amateur astronomer, I think it would make a light-weight, easily rotatable observatory cover. With a fabric cover, it should attain thermal equilibrium more quickly than a solid structure would.

Field Day Ham Shack: Put a coil-loaded antenna on top and use the dome's frame as it's ground plane and its base hubs for guy wire attach points! Wow, I think this is a no-brainer! I wonder how tall an antenna could be supported and what the radiation pattern would be like? CQ, CQ, CQ, DE KA7MWQ.

Scout Jamboree shelter/display kiosk. There must be some Merit Badge it would qualify towards!

Photographic lighting support or portable studio? You want a place to hang lights? You've got it!

Anything else you can think of?

Tools

At a minimum you'll need:

Tape measure, inch/metric of about 30 foot length.
Hacksaw w/metal blade(s)
Hammer
Screwdriver, x-point, #2
Marker pen, permanent, black, medium point.
Drill, electric
Drill bit, 3/16", stepped-tip, for metal*
Center punch
Protractor, 6" is nice.
Compass, pencil, for drawing circles.
File, metal
Hand saw, wood
Paper, blank
Pencil
Eraser
Straight edge such as a ruler, metric/inch

* Note: This bit worked very well and drilled all my strut holes. It has a short pilot drill bit at its tip combined with the final size of drill bit.

Contracted Work

Pre-cut 4.5" mild steel discs of approx 0.1" thickness, (22)
Pre-cut 4.0" mild steel discs of approx 0.1" thickness, (7)

Note: These discs will become the PENT (5-hole), HEX (6-hole) and BASE (4-hole) hubs for the struts. If you can take the time to cut them out with a saber saw from a large piece of mild sheet steel and drill the holes in 'em, go for it. But for $ 212.00 I had these hubs, including their 3/16" screw holes (152), fabricated by a job shop which had a Computer Numerically Controlled (CNC) X-Y table with a water jet cutting head. The precision is amazing and besides, what's your time worth? You'll be spending enough time making up the jigs, struts, and templates and then cutting and sewing the cover.

Tools I used:

    All the above tools plus:
    Vise, bench-mounted, approx 4" jaws.
    A drill press, bench-mounted, electric
    A hydraulic press, 20,000 lb, manual
    A saber-saw, electric, with metal and wood blades

The hydraulic press was used to squish the strut ends, the vise to hold the strut squished end to bend the strut ends and hold home-made jigs, and the drill press to drill the holes in the ends of the struts. If you don't have these tools, you could use the vise to squish and bend the struts' ends and the electric drill to bore the holes.

Tools to make my cover

My wife wouldn't let me within 5 feet of her fancy sewing machine and she wasn't sure if it would be able to sew the material I used for the cover, a white UV-resistant poly tarp material from a 24 foot square tarp found on the web. I didn't want to hire out the sewing of the gores to make the cover because it's just more money. So I did it myself... cut out the material and sewed it up... I'm pretty proud of that! I only broke one needle sewing approximately 150 feet of stitches. Came out OK. You can, too. I just made sure the machine could handle the material in multiple thicknesses with webbing strap in between.

     Sewing Machine, Kenmore Rotary Sewing Head, Model # 117.812, used, manufactured in 1948! Complete with owner's manual, Parts List, Buttonhole Attachment with its manual, and other miscellaneous attachments, used, $ 45.00! (Photo 3)
     Scissors (1 Pr)
     Pins, straight (50)
     Needles, sewing, packet of 5, suitable for the material
     Sewing Table, used, with cut-out for the sewing machine.
     3 other tables I scrounged from around the house to give me a fairly large support surface level with the sewing machine table for the material as I sewed.

Materials List

For some items I purchased more than I needed… for spares or fudge-factor in case I screwed up in measuring, counting, and for when I would surely loose a lock washer, etc., in the grass!

For the Frame:

    Hubs, Base, Hex, and Pent, (29) (contracted out)
    Electrical Conduit lengths, 10', ~3/4 inch O.D., 34 ea
    Nails and screws , scrap/scrounged for the jigs I built
    Lumber, scrap/scrounged for the jigs
    Plywood, scrap/scrounged for the jigs
    Screws, 10-24 x 1/2" length, x-point round head (140).
    Wing nuts, 10-24, (140)
    Lock washers for screws. (140)
    Screwdriver, stubby #2 cross-point

For the cover:

    Tarp, poly, UV-resistant, 24' by 24 ', "heavy duty", white
    Tarp, poly, UV-res., "heavy duty", silver, about 8' x 11'
    Thread, White, All Purpose, Polyester covered Polyester
        (!) 500 yds (1 spool)
    D-rings, metal, chromed, 1" wide, (12)
    Web strap, 1" wide, about 10 feet
    Clear, thickish flexible vinyl for 4 "windows".
    Tarp tape, 1 5/8" by 108-foot roll (2)
    Cordage, 1/4" poly, 200'
    Tent spikes, 15-20 ea., 8-10" long.

Design Considerations!

I designed my dome to be a temporary, waterproof (pretty much!), portable shelter. It would be covered in a light-weight, UV-resistant cover, have 4 "windows" ,two "doors", and a variable-opening vent up top. Components would be readily available and easily replaceable or repairable. For simplicity of form and it's use of minimal components while retaining a good "dome" shape, I chose the dome form known as a "2V icosa alternate". It consists of 2 differently sized triangles which make up its shape.

First question I asked myself was, "What materials would be adequate for the hubs and struts and available almost anywhere?" I had built three previous domes from plywood and 1" X 2" dimensional lumber. They were OK, but the hub design was weak and bulky and the wood split easily and would eventually rot. Also, the currently available 1" x 2" "furring strips" didn't look all that straight and knot-less at the local home supply stores. But the metal electrical conduit looked "appetizing". It was cheap enough and should be readily available everywhere in case I had to replace one. It's easy to drill and easy to crush the ends. I thought that mild steel plate would make an excellent and strong hub material, too. (Illus 1)

Next consideration: the diameter. I chose a diameter of 14 feet (center height of 7 feet!) because it seemed adequate for my purpose and would keep the length of its longer struts reasonable (no more than about 50") so they could support a significant weight in addition to the cover. I also wouldn't have to use a tall step ladder to get to the top levels. (The day I showed the assembled frame to my 5 and 7 year old grand-daughters, their eyes lit up. With a whoop and a holler, they were up on it before I could blink! It held 'em with one minor bend in one LONG strut on the side! Phew! I kept 'em off it from then on. Sure gave me heart palpitations, though! I straightened out the mild bend and it was good-as-new!)

Then I thought about the cover. I wanted it light weight, waterproof, translucent, and rot-resistant. Fabric wouldn't fill that bill; canvas or rip-stop nylon would be way too expensive and would rot. I knew blue tarp material was cheap but it never impressed me with its longevity. I was looking for something more translucent (white would be best), and a bit stronger. It would have to be UV-resistant, too. I found it "on line", a white tarp measuring, "before hemming", 24 feet square. Calculations showed it should cover the whole dome. I thought I might not have to make a hemispherical cover but just use the tarp tied down with two broom poles holding up a bit for a door and one pole holding up a vent opening opposite. (Photo 4) Unfortunately that turned out to be impractical.... strong breezes not only couldn't keep the temperature down inside without raising the whole tarp perimeter; the breezes played havoc with the loose areas of the tarp. Wow, the solar gain was intense. So I sewed a dome-conforming cover having a sizable door, a back door, and a 2 foot diameter top vent. A smaller tarp, silver, tied over the outside and facing the sun, provided for shade and temperature control in conjunction with the vent and doors. An even smaller triangular piece of white tarp scrap was made up to be used for the top vent area control. It's positioned by judicious placement of its three control lines, one from each corner to an appropriate place inside the dome.

Remember, this is all scalable. If you need the frame stronger or a larger diameter dome, you could increase the thickness and diameter of the hubs and use larger conduit and quarter inch bolts for instance. For whatever purpose and/or size you choose to build, have fun; learn a lot; and design and work safely!

The Jigs

The distances between strut holes and the diameter and placement of the holes in the hubs must be as precise as one can make 'em.

Enter the "jigs", a few templates, a compass, ruler, and drill punch (to keep the drill bit from "walking"). As construction of the dome proceeded, I made up jigs appropriate to the task at hand from scrap materials lying around my workroom and garage.

These simple jigs were made to construct the struts: one for cutting the overall lengths of the LONG and SHORT struts, a second for squishing the end of the struts, a third and fourth for drilling the holes in the struts, and a fifth for bending the tips of the struts to an approximate angle.

I had the hubs made for me but you can use a compass and drill punch to mark the mild plate steel, a jigsaw to cut out the hubs and an electric drill to drill their holes. It'll take a long time, but look at the bright side; it'll keep you out of the sporting goods and gun stores!

If you decide to make a soft material cover like I did, I've described and illustrated a template (pattern) I made to make an appropriately sized "gore" of which 10 were sewn up to make the dome's hemispherical cover. Another easy jig was made to mark their 1" seam widths.

Piece o' cake! Its fabrication just takes a little time and an old sewing machine and the guts to "put the pedal to the metal" (ask a seamstress).

Spreadsheets are cool!

Anyone, I repeat, anyone can learn to create a simple spreadsheet on a computer, even two-fingered typists. Although you don't need one to build a dome (just several blank sheets of paper), I created a few spreadsheets to help me build mine.

To learn how, you could start out with a making a simple check register or budget (you know you could use one!). Don't let the fact that it's a "spread sheet" intimidate you. Designing one to calculate the simple sums, divisions, and multiplications which the dome construction requires is easy.... the basic dome formulas are simple, just take it one step at a time. You can use a program's help files or maybe buy a "Spreadsheets for Dummies" type book.

The web site OpenOffice.org has a free, very powerful, "open source",and "cross-platform-compatible" software package you can download and use... it includes word processor, spreadsheet, database, presentation, and drawing modules. Great stuff! And since it's free, there's no mandatory financial cost to you, why not give it a whirl. (They'd appreciate a donation, though.)

For this dome, I tabulated my expenses on one sheet. On a second sheet, I tabulated and summed my labor hours and labor cost. Another sheet I created for calculating the dome's radius, circumference, diameter, strut length calculations, and hole-to-hole distances on the struts. A section of that sheet was also used to calculate the measurements required to make the conformal cover for the dome.

The neatest characteristic of a spreadsheet? You can play "What if...." scenarios such as "What if I wanted to use the maximum length of SHORT and LONG strut so as to minimize conduit waste. How big a dome would result?" Make one entry of the dome diameter, radius, circumference or even, say, the area you want it to cover, and, depending how you've designed the input and the calculations, the results are instantly generated. This book was written and printed with OpenOffice.org's "Writer", too, and many of the illustrations were done with TurboCAD, a very good and inexpensive CAD software package.

Metric Measurements

For those of you who are familiar with metric measurement, I say, "Good on ya!"

For those of you who haven't yet learned it, here's your chance. As far as linear measurements and building stuff goes, if I can use the metric system, I do. It is soooooo much easier to measure and manipulate the results. It is all decimal so conversion between meters (m), centimeters (cm) and millimeters (mm) is a breeze. No more worries about that last bit of length being 3/8" or 7/16" or worse, misreading those fractional marks between the inch marks on a tape measure marked in feet and inches. No more worries about adding and subtracting fractions of inches, inches, feet, and yards and converting between them.

But if you're used to that and comfortable with fractions, knock yourself out. For me, it's metrics and that's what I used for the construction of my dome. That's why I specified a metric tape measure in the tools list (You might have trouble finding one, a sad state of affairs!).

C'mon, get your feet wet in metric linear measurement! It's easy and takes no time at all to learn (and HEY!, no more fractions!)

```
1 meter (m) = 100 centimeters (cm)
1 meter (m) = 1000 millimeters (mm)
1 inch (in) = 25.4 mm or 2.54 cm
1 cm= 0.3937 in
1 ft= 0.3048 m
```

Suggestion: Make all your measurements and calculations in metric while building the dome and when you're through with the project, convert what's necessary back to the English system so others can "relate" to it. But you *will* end up with decimal inches which can be converted back to the nearest fractional equivalent. Very few rulers/tape measures are marked in decimal inches, if any!

Calculations Required:

I'd already decided on a 14 foot (4.267 m) diameter dome and my calculations for the other measurements required were based on this number. This style of dome, a "2V icosa alternate", requires 30 SHORT struts, 35 LONG struts, 10 BASE hubs, 10 Hex-Hubs, 6 PENT Hubs, 130 each of 10-24 screws, wing nuts, and lock washers. I made up or bought spares for all, two of each strut length, one of each hub style, and 10 extra nuts, screws, and lock washers.

In your mind or using a visual aid, picture a geodesic dome shape. It is composed of the joined apexes of triangles. Where these apexes join are just points in space. In place of each point will be a "hub" which would allow me to join the struts of the frame. This means the center of the hub would be very close to that apex point but could never get there because the hub would be flat and had a diameter. I designed my hubs for the smallest diameter possible allowed by the width and length of each to-be-squished strut end (which is the manor in which I chose to join struts to hubs). (Photo 8b)

The measurements I needed to calculate were the overall lengths of the two different struts (part of a chord, pronounced as "kord", a straight line segment which connects any two points on a curve). Fortunately, chord "factors" (multipliers) for the 2V alternate dome I was building had already been calculated. They are expressed as two factors of the radius. For the long chord, it's factor is 0.61803 and the short chord, 0.54653. So for my 2V alternate dome of 14' diameter, the long chord (strut) length required is 0.61803 * R (resultant radius of the dome) or 1.3186 meters (131.86 cm or 1318.6 mm – see? Just move the decimal point!) and the short chord length (strut) works out to be 1.1660 meters.

But there was a catch..... I had to adjust the overall strut lengths to take into account the bolt holes' distance away from the center of the hubs and the fact that SHORT struts attached to a P-hub and H-Hub but the LONG struts attached to two H-Hubs.

To help with the calculation(s), I drew up three diagrams to scale, one for each of the three different hubs and, not-to-scale, a diagram for each of the two lengths of strut. This way I could keep the fit of the struts to the hubs, their measurements, and calculations straight in my mind.

Designing The Hub(s)

I designed the hubs first, as I could then calculate the two lengths of strut required and their center-to-center hole spacing based on the hub design.

So I took a scrap piece of the chosen electrical conduit and squished one end of it in my vise. This gave me the width of the "squish". When bolted to either the P- or H-Hub, the squished ends of the struts must not interfere with each other and yet be as close to the center of the hub as possible for maximum strength of the hub-to-strut connections. (Photo 8b) To ensure that, I designed the two different hubs in the following manner.

I placed a dot in the center of an 8 ½ x 11 sheet of blank paper. For the P-Hub (Fig 1), using the ruler and protractor, I drew 5 radials evenly spaced (360 degrees / 5 = every 72 degrees) emanating from that dot. Taking that sample squished end I had made from a scrap piece of the conduit, I laid it between two of those radials on the paper until the corners of the squish just touched those radials. Then I slid the squished tip away from the center to allow for about 2 to 3 millimeters from each corner of the squish to the radials and marked that distance from the center. I drew a circle with the compass using that distance as the radius from the dot. This assured me that when the struts are bolted to the hubs, the struts' tips will not interfere with each other as long as they remained outside that radius. The measured radius of that circle was 23 mm for the P-Hub.

The maximum desired radius of the two different hubs would be determined by the length of the squish. For simplicity, I picked 15 mm from the tip of the squish to the center of its bolt hole and another 15 mm from the center of the bolt hole to the inner end of the squish (its flare). Now I had the additional radius I needed to figure the diameter of the P-Hubs. That would be 30 mm (15 + 15) plus the radius of that initially drawn circle (23 mm) which gave me 53 mm. I set the compass for a radius of 53 mm and drew a circle concentric with the center of the radials (the dot) and first circle.

From the dot, I marked a circle of 38 (23+15) mm radius. Where this circle intersects the radials are the centers of the screw holes. I chose 3/16" diameter holes. They'll take 10-24 screws with a little bit of slop.

I cleaned up my drawing and diagrammed my measurements, specifically, the distance from the center of my hub to the center of a bolt hole at 38 mm and the distance from the center of the hub to the outer circumference or its "radius" at 53 mm. I then labeled the drawing "PENT Hub". Under that label, I printed the thickness of the metal desired (about 0.1 " and "mild steel".

Keep in mind, if you use different materials, your measurements will differ. Check and double-check.

In a like manner, I created a diagram for the H-Hubs (Fig 2) except it had 6 radials, each at 60 degrees (360/6). For the hubs' and struts' ease of manufacture, I kept the allowance for the squish of the strut ends the same (total 30 mm) for all struts. For me, it worked out that from H-Hub center to bolt hole center was 43 mm (28+15) and so the radius of the H-Hub would be 58 (28+15+15) mm.

The BASE hubs (Fig 3) are H-Hubs cut off-center to allow for space to install the struts at ground level plus a little height off the ground. I allowed 0.5 inch (12.7 mm) and assured the cut was parallel to two opposite holes. The 12.7 mm is not critical, better more than less, but I felt that it would be a good minimum. It gives the 10 struts at ground level a little elevation off hard ground.

Designing the Struts

For the struts, I diagrammed them as follows. (Fig 4) Centered on a piece of paper, I used a ruler to draw a side view of a finished strut (with two squished ends and the screw holes). I wasn't attempting a scale drawing, just a crude representation. On paper it measured about 18 cm long. In the middle, I "broke" it with two parallel squiggly lines which tells an interpreter it's some length longer. On top of each squished end I drew a thin rectangle extending from where the squish begins on the strut out to the edge of the paper representing the side view of the two connected hubs, again, not to scale and showing the strut ends mounted on the hub.

Close to the edge of the paper and through the hub, I drew a vertical line to represent the center of each hub. And in the middle of where the hubs and squished ends overlapped, I drew a pair of vertical dashed lines about four mm apart down through the hub and squish area denoting the screw holes in the middle of the squish. I added a vertical solid line drawn halfway between those dashed lines to indicate the center of those "holes".

Now I could annotate my known measurements on the diagram and then figure the struts' overall length before squishing (e.g. how long to cut the conduit lengths) and the distance between the centers of the screw holes in the squish after squishing each end.

I came up with a SHORT strut (Fig 4) pre-squish cut length of 1.103 m and the SHORT strut holes' center to center distance of 1.085 m.

In a like manner, I drew out and figured the LONG cut-strut pre-squish length (Fig 5) and the LONG strut holes' center to center distances.

Remember, for the LONG strut, each end connects to a H-Hub. So my cut conduit length for the LONG strut worked out to 1.2632 m and its holes' center-to-center distance added up to 1.1772 m.

Making the Hubs

Caution!
During any fabrication, take all safety precautions necessary to avoid injury to yourself, property, or others!

As I stated before, I chose to contract out the manufacture of my hubs. The fabrication shop supplied the steel plate and cut out all the discs and drilled the holes. I didn't want to take the time. Their $ 212 charge for the time and materials was well worth it.

If you don't want to job it out, I think that, as a minimum, you'd not want to deal with a large sheet of heavy, floppy steel to cut out the discs for the hubs. However I do offer some opinions on the matter...

One could cost out a job shop cutting out all the hubs as I did.... it might be cheaper in your area (Got any friends with a CNC water/plasma/laser cutter? How's about asking your local high school, or area vocational school as an exercise for training their students?)

One could have the shop cut out strips of the sheet steel from which you'd then cut out the hubs yourself with a saber saw. Diagrams will help you plan out the dimensions of the strips.. don't forget to allow for an additional spare H-, P-, and B-hub.

To make the cuts, perhaps you could rent or borrow the use of a metal band saw. If you have a saber saw, fine. I'd clamp the workpiece, for sure, and have a fresh packet of blades handy.

For the drilling of the holes, you could make a hard template from your drawing and center-punch all the holes, then drill a small pilot hole at each, and follow that with a 3/16" drill.

For repetitive fabrications such as this, I'd perform a single operation on all pieces.... first, I'd layout the pattern on all the plates and double check 'em. Second, I'd center-punch all hole centers on all the plates. Next, I'd drill all the pilot holes, and finally, I'd drill all the 3/16" holes. This would save time by minimizing set-up and take-down to perform the different operations.

I'd also take a file and smooth off all edges and hole perimeters on both sides so no one can get cut.

Making the struts

Making the struts consists of 8 operations: 1) cutting the struts to pre-squish length, 2) marking the length of squish on each end of all the struts, 3) squishing the ends, 4) de-burring (filing) the ends, 5) drilling the first hole in each strut, 6) drilling the opposite hole in each strut, 7) Cleaning out the holes, and 8) bending the strut tips. All of these operations beg one to create jigs to assure identical lengths, hole positioning and hole spacing, which is critical.

Operation 1) To cut the struts to pre-squish length, I made up a jig (Fig 6) consisting of a long board, a few short lengths of 1 x 2, a small piece of angle aluminum stock, and screws, all "found" items.

First, I made the cutting end of the jig to ensure the jigsaw would ride smoothly and level over the end of the conduit and ride against a guide so the jigsaw would always cut perpendicular to the conduit at the same length. I also made sure that the blocks the jigsaw rode on and which also straddled the conduit to keep it from moving, were not so high as to prevent the blade from cutting all the way through the conduit.

Then, from the cutting plane of the blade, I measured back to the length required for the SHORT strut and affixed the angle aluminum with two screws (could be just another piece of wood) to automatically position the conduit at the correct length of cut. To keep the conduit centered on the aluminum backstop, I mounted two pieces of wood on either side of where the conduit would rest and just in front of the backstop so I could just slide in the conduit, brace it against the backstop, and then run the jigsaw through at the cutting end.

I made a test cut and ensured the length was correct. Adjusted as required. Fed the next length in, cut, and repeated for the number of SHORT struts required (32).

For the LONG strut pre-squish cuts, I affixed the backstop the appropriate distance further back; verified the first cut length; adjusted if necessary, and proceeded for the remainder of the LONG struts (37 total).

Operation 2) Squishing the ends was the next step. Whether or not you use a hydraulic press, a large vise, or a hammer, one still has to mark the 30 mm squish length position on each strut's ends. One can exceed that length by a few mm, but make sure it's at least 30 mm.

The purpose of the jig would be to mark every strut end with a line 30 mm from each end. To mark that distance consistently, I made a jig similar to the one I describe herein.

This jig (Fig 7) consisted of a small piece of wood 30 mm thick and about 10 cm x 4 cm nailed/screwed/glued (whatever!) to a base of any other conveniently sized piece of wood. The 10 x 4 piece has a notch carved, cut, drilled, or chiseled in it to allow the vertically-held conduit's end to rest inside it on the base. Facing the notch is another rectangular piece, about 30 mm thick, which will trap the end of the conduit in the notch yet still let it rotate. After building it, rest the tip of the marker pen on the top surface of the 30 mm thick wood and while lightly pressing the tip against the conduit, rotate the conduit, thus marking the end.

Haven't got or can't make up a 30 mm thick piece of wood? You could use a nail or screw head as the pen tip's reference by mounting the nail in the apex of the notch and sinking it to a height of 30 mm above the base wood surface. Rest the pen tip on that while turning the conduit.

Mark all conduit ends 30 mm from the end.

Operation 3) Squishing the ends....CAUTION: **The ends of each strut must be squished in the same plane.** The first end squish of each strut with my hydraulic press was made without concern, but the opposite end's squish had to be IN THE SAME PLANE AS THE FIRST! To ensure this (Fig 8), I set up a rest which was parallel to and level with the plane of the press's anvil at just under the length of the strut away from the anvil of the press. The already-squished end would lie on this rest which causes the other end to be squished parallel to it. (I made sure the strut didn't rotate as I squished). With my manual hydraulic press, I inserted the tip of each strut between a pair of steel blocks (came with the press) up to and just covering the 30 mm mark, laid its other squished end on the rest to assure the resultant squish would be in the same plane and then pumped like mad until the end was squished. I checked the length of "squish" on the first few squishes to make sure I had at least 30 mm of squish length, then squished 'em all, first squish, then rotate and lay the first squish on the rest, second squish; next strut!

Without a drill press, I'd insert the ends of the struts, one at a time, into a vise to the marked depth (having the vise jaws cover the mark completely to give me a couple of mm more "fudge factor") and turn the handle to squish the first end so no gap exists at the tip of the strut. I'd verify the results on the first strut, adjust if necessary, and proceed. For the other end of each strut, I'd set up a vertical jig giving a rest parallel to and above the jaws giving the already squished end its rest so I could assure the new squish occurs in the same plane. Push comes to shove, I suppose you could just hold or clamp a foot-long flat, thin, and straight piece of wood to the already-squished end; line it up parallel to the vise's jaws, and squish the other end.

No vise? Perhaps I'd make a manual press in the style of a lobster or crab cracker with 4 foot (or so) arms. One arm would rest on the floor, and the conduit end would be squished close to the pivot, thus giving me significant leverage. Improvise!

If all else fails, I'd pound 'em flat with a hammer by placing the strut end between two hard pieces of wood and banging away. Just make sure the squishes end up being in the same plane and of the minimum length.

Operation 4) Filing the ends of the struts.... This took off all the raw-cut edges at the tips of the struts. One could get a pretty vicious cut by inadvertently brushing ones skin against 'em so I was wearing work gloves as I filed away the rough stuff.

Operation 5) Drilling the *first* hole in each of the struts... I used a jig (Fig 9) bolted to the drill press platform which placed the drill bit's position centered on each strut and 15 mm in from its tip. Edge to edge centering of the hole is not that critical and one needs some play to insert the strut tip in the jig, allowing for squish width variations. I would nudge the tip into the same corner of the jig every time just for consistency. I put one hole in each strut in this manor (69 holes).

Operation 6) Drilling the other hole in each strut... Here's when one wants to take the utmost care and design the jig to give the most accurate results. Correct distance between the hole centers on each strut is extremely important. Having my drill press and vise mounted on a common surface and fairly close to each other, I felt the most accurate way to set the holes' center-to-center distance would be to set up the center of a fixed pin of 3/16" diameter to (as physically accurate as possible) be at the correct distance from the drill press drill bit's center position. In this manner I used the pin as an index so that where the drill bit dropped to the drill press's table would always be the correct distance from the hole on the other end for every strut.

My jig (Fig 10) for drilling the struts' 2nd hole consisted of a horizontal base (think 1 X 4 on edge) which allowed the vise to grip it and which was long enough to permit a linear adjustment towards or away from the drill press and up or down enough to set a pin's (another 3/16" drill bit or an upside down screw) height in relation to the drill press's platform.

I affixed a platform to the top edge of the base at the farthest end from the drill press to hold a vertical 3/16" drill bit or 3/16" screw protruding about 1 cm above the surface.

The assembly was then clamped in the vise to set the pin at the same level with respect to the drill press's platform. At the same time, I set the distance from the center of the drill bit to the center of the pin (1.166 m for the LONG struts) while maintaining that height. Some index marks and lines placed on the vise and the base of the jig helped considerably.

The process for drilling the last hole in each LONG strut ran like this... I placed the strut's first-drilled hole onto the pin and, with the other end of the strut placed into the jig on the drill press platform, I drilled the strut's other 3/16" hole. (Note: I re-positioned the drill's platform jig a few mm away from the end of the strut to allow the squish to fall into the jig under the bit yet still be centered edge-to-edge). After drilling the first strut, I measured the distance between the drilled holes with the tape measure to verify the jig's setup. If you're not confident of estimating where the center of the holes are, and since they are the same diameter, you can measure from right edge to right edge or left edge to left edge (1.66 m or 166 cm for my long struts). This process guaranteed that the distance center-to-center between the holes on each strut would be correct. I did this for all the LONG struts. Then I reset the vice mounted jig for the proper pin distance (1.1772 m or 117.72 cm) for my SHORT struts and finished the remainder.

Without a drill press.... If I hadn't had the drill press, but just an electric drill, I would have set up a different jig (Fig 11) such as a wooden compass with a base about 1.8 meters long, a 3/16" or 4 mm pin or inverted 3/16" diameter screw at one end and a fine-tipped pen or metal scribe affixed at the appropriate distance at the other end.

I'd center punch, pilot drill, and drill the first hole in the middle of one squish area of each strut as accurately as I could. You could make a jig for that to mark the centered position.

After drilling the first hole in the center of one squish of all the struts, I'd scribe 'em all using the Fig 11 jig set up for the appropriate length of strut, and then center-punch and drill the second hole on the scribe mark midway between the edges of the squish at the other end. The drill punch should help prevent the bit from "walking" while starting the hole. Again, check and double check your first few pieces before you make the "run". Measure and adjust as required. Proceed for the remainder of the struts.

Operation 7) Cleaning out (de-burring) the drill holes. I used a manually turned 1/2" drill bit to take out any burrs in the holes.

Operation 8) Bending the strut tips.... The struts will be bolted to the inside-facing surface of each hub. That way the wing nuts won't tear up or snag any cover lying on the dome frame and the round screw heads would be less than likely to chafe against the cover. At this point in the fabrication, if I were to bolt three LONG struts to two BASE hubs for the ground connections, they'd form a pretty good straight line. I needed to make it so the struts mate with the hubs at an angle so when all bolted up, they form a circle of the correct diameter with minimal stress. There's an easy way to determine the approximate strut-squish to strut angle to set.

On a blank sheet of paper, (Fig 12) I drew a circle of radius 3.5". I drew a diameter line and then marked a radial at 36 and 72 degrees. Where each of these intersects the circumference, draw a tangent (a line intersecting and perpendicular to the radial at the circumference). These tangent lines represent the plane of the BASE Hubs. Next I drew chords which intersect each radial (representing a couple of struts connecting the BASE hubs. I then measured the angle between the strut and BASE hub line previously drawn and came up with about 18 degrees.

I realized that this angle would be different between P- and H-Hubs, figuring the P-Hubs would require more angle. But I bent all to the one angle in the interest of my sanity and simplicity. It seemed to work alright.

To bend the strut tips to the approximate angle, I made up another jig (Fig 13). This jig rested off one end of the jaws of the vise and had a vertical line beginning at the closed jaws of the vise extending upwards an appropriate distance scribed upon it. From that line's origination at the top of the closed jaws, I drew another line upward at an angle of 18 degrees from the vertical line. This gave me a reference to which I bent the strut. I then bent all strut squishes to this angle by clamping the squish area completely into the vise and bending the strut to the 18 degree line. The process required that the strut be pulled some small distance beyond the 18 degree line so that when released, it would rebound to the line. I didn't fuss over getting it exact, though.

And THAT'S IT! I now had the components built for the frame. As soon as I could, I assembled the dome.

Before Assembly

Before I hauled all the hubs, struts, wing nuts, bolts, and screws, etc. out to the yard, I made sure I had an area 15 feet in diameter clear and flat. It didn't have to be level. Or, I figured, I could always shim the BASE hubs with rocks or wood where appropriate once I assembled the dome, if necessary.

To mark the position of the dome's base, I made a compass out of a tent peg with a cord length equal to the dome's radius then traced out the dome's circumference. I marked it's path using the BASE hubs.

After I assembled it the first time, I found an empty kitty litter bucket to hold all the "stuff" to assemble the dome, e.g. cordage, rope, hardware, tent pegs (now dome pegs), tarp tape, compass, and assembly instruction, and a coffee can for the hubs.

When I assembled this dome the first time, all went smoothly except for the final tightening of the wing-nuts. Tightening those wing nuts was hard on my fingers. And once tightened, loosening them on take-down also was difficult for the first turn or so.

Rather than lugging around a crescent wrench with the dome, I built a tool (Fig 14, Photo 5) from a scrounged section of a wood tool handle (could have used a 2x2 piece, too) to aid in this function.

I cut a section about 4" long (whatever fits your hand) and flat on one end. On the flat end, I drilled a centered 3/16" diameter hole about 12 mm deep. I followed that up by drilling a hole in its center with a width a little more than the diameter of the round perimeter of the wing nut, about 1/2", and 4 mm deep. I then took a saw and made two centered and parallel saw cuts across the surface, 9 mm deep, parallel, and passing through the center, spaced opposite edge of cut to opposite edge of cut and about 5 mm apart to take the "wings" of the nut. I cleaned out the area between the cuts and checked to ensure that a wing nut would nestle in and rest flush with or a little deeper than the surface. The result was a self-centering tool to place over the wing nuts and apply adequate torque with my whole hand. It sure was easier than using my fingers. I stored it in the kitty litter pail.

The only improvements to the tool I made were 1) screwing in an eyelet at the other end for a wrist cord then attaching the cord (scrap leather shoe lace) with a large fishing swivel between the cord and eyelet.. and 2) roughing up the tool's hand surface for more grip friction. I stood the tool on its uncut end and clamped it up in the vise, I drew a saw's blade across the surface while raising it in-line with it's axis. I made these strokes all around the perimeter, then reversed the direction of the draw of the saw on the handle and did it again. The result is a shallow checkering pattern all around the handle. I cleaned off all splinters. (Note to self: maybe add a magnet deep into the tool's throat to retain a wing nut and lock washer)

I also needed a way to have on my person while I assembled the dome, the wing nuts, bolts, and lock washers. I first chose to use a large plastic coffee can which I hung from my belt loop on my left side. That way I could just dip in where I was instead of moving to a can on the ground, where ever it happened to be. That was a back-saver at my age.

The only improvement I thought of for this carry method was to compartmentalize the lock washers from the wing nuts and bolts. The lock washers always seemed to bury themselves at the bottom of the can under the wing nuts and bolts, which made it very difficult to pick them out for the next connection. So I found a suitable container (a powdered edible fiber container... I told you I was getting' old!), and attached it to the outside of the coffee can with nylon ties. This way I kept the lock washers separate and easy to pick. The containers were left deep to prevent the hardware from falling out while moving about (Photo 7).

I'm not a tall person, standing all of 5' 5" if I'm lucky. Guess what I needed to assemble the top two levels? A folding stool. I bought a small collapsible one at a big-box retailer. Helps me reach the middle of the windshield on my Jeep Laredo, too, to clear the ice in the winter! (I suppose the kitty litter bucket could do double-duty! Just be real careful!)

Assembly

When I set up this dome, I relied on my old instructions I had written in March of 1981 for the three wooden domes! They worked fine but I updated them here for this metal-framed dome:

General assembly practices: I kept the nut, bolt, lock washer combinations loose (just enough to fully thread the wing nut on the screw to where the tip of the screw's threads was flush with the wing nut and not protruding from the wing nut). I didn't tighten them until all connections were made and the hubs set "upright", (not rotated off kilter.... you'll know it when you see it. Photos 8a and 8b). Then I used the custom-made tool and a #2 cross-point stubby screwdriver (Photo 5) to snug 'em up (which I've also stored in the kitty-litter pail).

After bolting free struts to hubs, ease them down to their natural resting place if they want to drop. Until the dome is completely bolted up, there could be some major stress at the hub/strut junctions so try to minimize that stress.

All struts and wing nuts should be mounted on the inside-facing surface of the hubs. This will minimize any cover chafing caused by the wind or snagging and tearing as it's pulled over the frame.

Most assembly is done from the inside of the dome.

Occasionally you might have to tug or push a little to align screw holes of hub and strut. No worries, mate! (As long as you've cut the struts and holes correctly!)

While assembling or disassembling struts to hubs above ground, I rested the strut on/against my shoulder to support it while I screwed 'em together.

FIRST LEVEL

Illus 2. I put down the 10 BASE hubs evenly spaced and along the circumference as stated in the previous chapter. I picked one BASE hub and called it (labeled it) #1. (I could have marked it with a tent peg beside it or something.) I put it where I wanted the front door.

Illus 3. I laid out end to end 10 LONG struts on the ground between the BASE hubs to form the ground level approximate circumference of the dome.

Illus 4. I then bolted those struts to their respective hubs using the two holes closest to the hub's straight edge and on what would end up as the inner facing side of each hub when the BASE struts are all bolted up in a circle. I made sure I bolted the struts so their tips curled in the direction of the circumference.

At BASE hub #1, I laid a LONG strut's end against the hub's left edge and pointing towards the center of the circle.

On the next left (CCW since I stood inside the circumference) BASE hub (#2), I laid another LONG strut end against that hub's *right* edge and pointing towards the center. The two LONG struts were now "neighbors" between the two BASE hubs, #1 and #2.

I picked two SHORT struts and similarly laid one against the left edge of #2 BASE Hub and the other against the right edge of the #3 BASE hub.

Illus 5. In a similar manner, I alternated pairs of LONG and SHORT struts around the circumference. I double-checked their placement and lengths... two LONG, two SHORT, two LONG, two SHORT.... etc.

I bolted those struts to their respective BASE hubs, letting them lie on the ground. Again, I double-checked to make sure that I had alternating pairs of LONG and SHORT struts. That would be a LONG and a SHORT on each hub but so arranged that a LONG faces a LONG on the neighboring hub and a SHORT faces a SHORT on its neighboring hub.

Illus 6. Then, I carefully raised a pair of SHORT struts to form an inverted "V" and bolted their ends to adjacent holes on a P-Hub. Use other struts to brace the structure as you go along until it's self-supporting.

Illus 7. I next raised the neighboring pair of LONG struts to form an inverted "V" and bolted their ends to a pair of adjacent holes on a H-Hub.

Illus 8. Between those two raised hubs, I bolted a SHORT strut. I continued around the structure until I completed the first level. This gave me alternating P- and H-Hubs forming a complete circle, the first level.

SECOND LEVEL

For the second level, I grabbed a H-Hub, two LONG struts and one SHORT strut. I bolted the SHORT strut to one hole of the H-Hub and on either side, I bolted a LONG strut as in Illus 9. I made sure they were on the same side of the hub (this is the "inside"). Likewise, I assembled four more H-Hub "sets", resting each assembly outside and against a P-Hub on the first level, their insides facing the center.

Next, I mounted an assembly by taking the end of a left LONG strut (I'm inside the frame) and bolting it to an H-Hub on the frame, then the center (SHORT) strut to the clockwise P-

Hub on the frame, and finally, the right LONG strut of the assembly to the next clockwise H-Hub. The assembly should stay up without support, now. Illus 10

Once I had all five HEX assemblies mounted, I joined the top H-Hubs with a LONG strut forming a top pentagon. And that's the Second Level done! (Ill 11)

TOP LEVEL

Now for the "Grand Finale"! I took three SHORT struts and bolted one of their ends to a P-Hub while on the ground.

Then I raised the assembly and rested it on the top of the frame such that the three free strut ends rested near their respective H-Hub. I bolted them up. Then I bolted up the last two struts. (Ill 12)

And Lastly!

I examined the rotational position of each hub and if it was properly oriented, I tightened its wing nuts with the home-made tool and #2 stubby screwdriver. If not, I'd rotate the hub to its correct position and then tighten the wing nuts. Photo 8a (incorrect) and 8b (correct).

Then I made a final check around at all the hubs, checking and tightening the wing nuts that I'd missed. If there were any significant gaps between the BASE hubs and the ground, I shimmed 'em. It was a lot easier going the second time I erected the dome, believe me!

Photo 6. The kitty-litter pail and coffee can in which I stow the dome parts (except the struts!)

Photo 7. The coffee can (for screws and wing-nuts)mated with a fiber suplement container (for the lock washers) and a hook through the lip of the coffee can to hang 'em onto my belt loop.

Cover Up!

By far, the best cover would be a dome-shaped skin to cover the outside of the frame. I had tried to use just a 24' X 24' white poly tarp, but any breeze would toy with the tarp and, in the evening, the noise would, as I found out in the high desert of south central Oregon, prevent me from going to sleep. I had to resort to wearing earplugs until I got a properly shaped (hemispherical) cover on it. What a difference!

Since I already had the 24' X 24' tarp, I used TurboCAD to map out 10 gores to make sure I could use the tarp as a source for the cover material. It would be a 10-gore hemispherical shape with a 2" skirt. The ground edge (bottom) of the cover would use the two longest grommeted edges of the tarp as the skirt so I wouldn't have to install the grommets. The tarp did prove large enough that I didn't have to sew panels together from a roll of material to make up each gore width.

To securely hold the cover on or over the dome so the wind couldn't blow it off and to minimize wind-caused noise, I decided to place 10 tie-downs (Photo 9) equally spaced around the inside of the dome cover several inches above the ground level and centered on each gore. This way I could either tie the cover to the frame of the dome or to ground pegs (or both) without going outside.

I decided to have two doors opposite each other for good cross ventilation and egress/ingress (cover photo) and placed such that I could remove one strut at the front door position to allow for a large door opening without having to stoop when entering or exiting the dome. The doors would be cut out so as to be the shape of an opening in the struts with the one strut removed. They would be (Photo 10) covered or closed with a flap pivoting along one upper border and sewn to the inside of the cover. They would have a tie-down sewn opposite to the seam (Photo 11) and be easily secured from inside or out.

I wanted 4 small windows, also, to allow those inside a 360 degree view of what could be going on around the dome, one in each door flap (Photo 12) and two others, each about 90 degrees from the doors placed at eye level (Photo 13) and so as not to be crossed by a strut. I tried to sew these windows onto the tarp material, but the stuff just wouldn't allow it. So I ended up taping the transparent plastic material with white 1 5/8 inch-inch "tarp tape".

I made a triangular vent cover (Photo 14) out of leftover tarp material which I was reluctant to discard. So far, I've made a printer cover from scraps and am going to make a scanner cover, too.

The sequence of making the cover went like this: made a half-gore template; traced the gores; cut out the gores; made and sewed 10 tie-downs; suspended two gores and traced the door openings; suspended two more gores and traced two window openings; cut out and taped the door openings' perimeter; cut out the door flaps; cut out and taped in the doors' windows; cut out the other two window openings and taped them in; sewed the tie-downs onto the door flaps; sewed the door flaps to the door gores; marked the seam allowance on the gores; and finally, sewed the gores together. Phew!

Making the Half-Gore Template

First, I made a template of half a gore (Photo 15) to facilitate me tracing out each gore on the tarp. Using a half-gore template meant I needed only half as much material to assemble the template and it would be half as heavy and easier to maneuver and store if I needed to make another envelope.

To calculate (spreadsheet anyone?) the template dimensions, I measured the required diameter (D = 4.267 m) of the cover (the widest), calculated it's circumference (C = 3.14*D,

13.4 m), decided on how many gores (N) I wanted (N = 10) and divided C by N, 13.4/10, which gave me the width of each Gore (W = 1.34 m).

I then calculated the length (height?) of each gore by dividing the dome circumference by 4 (L= 13.4/4 = 3.35 m). But remember, I had to also account for the skirt height (51 mm).

For my template I scrounged and pop-riveted up some thin fibrous material (looked like dry-erase board and was stiffer than cardboard) to measure L (3.35 m + .051 m [skirt] = 3.401 m) by W (1.34/2 m + .0254 m seam allowance = 69.54 cm) because the template will be half the width of a gore including its seam allowance along the gore's curved edge.

Next, I marked the "ground level reference" with a line 51 mm above and parallel to the base edge (which would be a grommeted edge of the tarp material). This would be the start of the following division marks. I then drew a line every one-tenth the distance between the top and the ground level reference line of that template, edge to edge and parallel to its ground level reference line. (Ill G1)

From the left vertical edge of the template, and at the ground level reference line, I marked a first point on that ground reference line 1/2 of the gore width (=67 cm) +2.54 cm [seam allowance] away from that left edge. (Ill G2)

At the next line above, I placed a mark in a like manner only (.4938 * 134 cm)+ 2.54 cm from that left edge.

At the next line above, I placed a mark in a like manner only (.4755 *134 cm)+ 2.54 cm from that left edge.

At the next line above, I placed a mark in a like manner only (.4450 *134 cm)+ 2.54 cm from that left edge.

At the next line above, I placed a mark in a like manner only (.4045 * 134 cm)+ 2.54 cm from that left edge.

At the next line above, I placed a mark in a like manner only (.3536 *134 cm)+ 2.54 cm from that left edge.

At the next line above, I placed a mark in a like manner only (.2939 * 134 cm)+ 2.54 cm from that left edge.

At the next line above, I placed a mark in a like manner only (.2270 * 134 cm)+ 2.54 cm from that left edge.

At the next line above, I placed a mark in a like manner only (.1545 *134 cm)+ 2.54 cm from that left edge.

At the next line above, I placed a mark in a like manner only (.0782 *134 cm)+ 2.54 cm from that left edge.

I placed a mark 2.54 cm in from the left edge at the top. The top left corner of that template now is the apex of the half-gore panel. (Ill G3)

I then drew the smoothest curve I could through those points from top to bottom and cut off the excess of the template outside the marked curve to give me a template for half of a gore plus the seam allowance along the curved edge.

Now I had to decide how large a vent I wanted in the top of the cover. I settled on a 24" (61 cm) diameter hole, figuring that if needed, I'd have enough room to stick a stove or fireplace flue pipe through it to vent smoke. Heck, maybe I wouldn't need a flue pipe, but the dome would definitely require the vent.

Using a marker with a string tied onto it close to its tip, I traced an arc of radius 12" with its center at the apex on the vertical edge of the template across to the curved edge of the template. I then cut that top off. Once sewn up, the results should be a 24" diameter hole at the top of the cover. (Ill G4)

Tracing & Cutting Out the Gores

I made sure the tarp was spread out on a flat surface and that the tarp stayed fairly flat as I traced on it.

I placed the base of the template on the tarp material with it's curved edge against one straight edge and corner of the tarp and it's base aligned with and on the tarp's perpendicular edge with the grommets. I then traced the template's curved edge on the tarp and the top arc. I marked the inside base corner and the right end of the traced arc at the top for references.

I then rotated the template over on its long straight edge to trace out the other half of the gore, making sure the top and bottom were aligned with the top and bottom reference marks made above, and traced the opposite half of the gore.

I reviewed and confirmed my measurements, markings, and tracings and then cut out the first full gore panel. This gore I suspended and clamped over the assembled dome frame from top to base to check that it fit.... seemed alright. I measured its base length, subtracted 5.1 cm, multiplied that times 10 (10 gores) and compared that with the required circumference. Seemed OK, it came out a little bit more. So I traced and cut out the remainder of the gores.

As my Latin teacher of long ago said, "Alia iacta est!"... in English? "The die is cast!"

In the center at the base of each gore, I numbered them 1 through 10 on what would be the inside surface. #1 gore I used for the front door and # 6 gore for the back door. Numbering the gores allowed me to keep them straight in my mind and correct in their assembly order.

Since sewing the 10 gores together would be progressively more ungainly and then attempting to sew the doors and tie-downs to the cover would be damned near impossible, I made and sewed on the tie-downs first; I put the windows in the door flaps next; put the other two windows in their respective gores (#s 4 and 8), then sewed the the door flaps on their gores. Only after those operations were complete, did I sew all the gores together to make the cover whole.

Making and Sewing the tie-downs

So I asked myself, "Self? How can you attach a tie-down to this relatively flimsy tarp material? It's gotta be strong and last a long time. It has to spread the force of a rope pulling on the material so the force doesn't easily rip the attachment point at the material. It's gotta be cheap and easy to make, too!"

Drawing on my continuing fascination with blimps and dirigibles, I thought of the "catenary curtain". On blimps, this is how they "attached" the cabin to the envelope. The cabin hung from a catenary curtain inside and attached to the inside surface of the top of the envelope. Something similar could be made with a D-ring (the cabin) and a length of web strap and scrap tarp material (the catenary curtain). The wide top of this assembly could be sewn to the material thus spreading out the forces acting upon the D-ring.

Making up ten of these attachment points would be a simple introduction to and familiarization with one's sewing machine if need be. Worked for me!

But first, I had to design one of 'em. Hmmmmm. PATTERN TIME!!

To illustrate what I came up with, take a blank 8.5 X 11 sheet of paper and fold it in half along its width. Open it back up and fold the sheet in half along its length and again, open it up. This gives you another center line for reference and where the folds intersect, the center of the sheet. Draw a centered 1 inch line along the width's crease. Along the top and bottom edge of the sheet mark off a centered 6 " length. Connect one end of the central 1" line to the two marks on each same-side top and bottom edge. Do the same for the other end of the central 1" line. You should have a bow-tie shape outlined. Cut it out with the 1" line being the center of the "bow-tie". There's my pattern.

After cutting out twelve of these "bow-ties" from excess tarp material, I then cut twelve 6" lengths of 1" wide webbing.

Then I slipped the D-ring over the bow-tie to the center, threaded the strap length through the D-ring half way and between the d-ring and material, folded it over, and spread it out in a V shape so it lay along and just inside the edge of the bow-tie of material. I then folded over the free end of the bow-tie at its narrowest width to give me a triangle supporting the D-ring at its apex.

I then sewed the assembly by first sewing across and just above the D-ring to fix the D-ring and web strap in place while paying attention to the V spread of the web. I sewed two seams on the strap length-wise and fairly close to the inner and outer edge of the strap legs. I ran a stitch all the way across the top ends of the V and then another two seams across the top of the tie-down (bottom "edge" being the D-ring). This top edge will later be sewn to the center of each gore about a foot above the gore's ground reference line. (Photo 9) Only nine more to go!

If you're building one of these covers, I suggest you number your gores along the base in the center on what will be their inside from 1 to 10 starting with the front door gore. Numbering them will help you sew the gores together in their proper sequence and will tell you on which gore you're cutting/sewing. Once you start sewing the gores together, you don't want to have to rip out all them stitches!

Gore Doors

I suspended two gores opposite each other and traced the door openings a couple of inches inside the struts which outlined the door position as I wanted some overlap of the struts by the door opening. (Photo 16) I cut out and taped the two door openings' edges. This should prevent a possible tear of the material if someone should sideswipe it while entering or exiting the dome.

I cut out the door flaps from left over material, making sure that the flaps were 3 inches wider and taller than the opening and roughly the same shape. This gave me sufficient material to sew the flap to the skin above a top edge of the opening and since it would be inside the skin but outside the frame, it would be held against the skin by the struts when closed and when opened, held flat beside the door frame by the neighboring struts. (Photo 17). Opposite the to-be-sewn edge of the door flaps, I sewed and taped a web loop directly to the doors so I could tie the flaps down or to the side and in what would be the top half of the panels, I cut out the window holes and taped in the window panes.

Gore Windows

I suspended two more gores opposite each other where I wanted the windows and traced two openings in their desired positions where they would be at roughly eyeball height (for me, anyway) being careful that struts wouldn't interfere with them. (Photo 13). Better too low than too high. I cut out the holes and taped in the clear plastic material.

I sewed a D-ring-less tie-down onto the inside of each door flap then sewed the door flaps to the door gores and taped over the seam (Photos 11, 17); marked the seam allowance on all the gores' edges with another jig (Illus 13); and finally, sewed the gores together, being careful to make sure the doors and the windows would line up opposite each other (that's why I numbered them). Hint – I sewed 5 pairs of gores together, then sewed those 5 pairs to each other. This is when the numbering of the gores kept me sane. I had to make sure that the gores were sewn so they all had their numbered surface INSIDE and that whenever I put two gores together with pins prior to sewing them, I had what would be their outside surfaces face to face. Then I could sew them on a line half way into the seam allowance. Got it? And, yes, I stuck myself several times!

Once I got all the gores sewn together, I taped the vent hole's perimeter, too.

Covering the Dome

To cover the frame with the skin, I needed a length of rope or heavy twine and a broomstick or broom with no sharp edges. Once I assembled the dome's frame where I wanted it, I laid out the envelope next to the dome, inside out, folded in half, and oriented with the vent hole closest to the dome and such that when it's finally on, the doors would be roughly where I wanted them... between the proper struts. I also had to take into account the prevailing wind's direction and strength and attempt to figure out how it could aid or hinder the operation. Again, the numbers on the panels helped and so did the # 1 on one of the base hubs... since, when I assembled the frame, I took care to position that BASE hub where I wanted the front door to be.

    I got a scrap piece of 2" PVC pipe about three feet long and inserted it into the vent hole and then tied one end of a long rope to its center. Then I threw the free end of the rope over the dome frame and used it to gently raise the vent area to the top of the dome. If anything hung up, I would use a broomstick/broom/hand to free it and then haul it up some more till it reached the top. I then drew the cover's "ends" around so I had half the dome covered. Next, removing the pipe from the vent hole and using the rope, I attached the rope to the D-ring at the base of the front door gore and gently hauled the base of the cover up, over, and down to cover the remaining half of the hemisphere. Using my hands and the broom, I got the envelope oriented just right.

I secured the cover and frame as I thought necessary. If I kept the vent closed in the summer, the sun would suck the moisture out of the ground and it'd collect on the inside of the skin. It also got awfully hot inside. I found the gray tarp can shade the dome somewhat if it's suspended reflective side out and facing the sun. It had to be adjusted several times to track the sun, but with the doors and the vent open, I found it to be tolerable even in summer, especially if I raised the skirt, giving me maximum ventilation. Once it cooled down in the eve, I'd close it back up. I would be sure to vent it with any size of open fire inside....!!!

Has it been worth the expense and effort to me? You bet! If you build one, I'm sure you'll enjoy it's utility, versatility, and comfort! Don't be afraid to try your own improvements or modifications. "They" say, "Necessity is the mother of invention", so design and build it as *YOU* want it. Just build and use it with safety as your prime consideration.

Author's Biography

Tom Newbery likes to think of himself as a displaced Connecticut Yankee who currently lives in Utah with his wonderful wife of 42 years, two cats, and, according to my wife, way too much "stuff". (But you ought to see her "sewing room"... it takes up one floor of our home!)

He began his work career as a florist's delivery boy at the age of 15, then was a waiter and "second" cook at a family summer camp on Mt. Desert Island, Me. He graduated from "prep" school (barely) and began his freshman college studies in Boston, Ma., studying electronics. He got discouraged with that (studying, not electronics!) and ran off to Sebring, Florida, where he worked as a bus-boy. Fortuitously, he had purchased a used 1948 Plymouth car at Happy Harold's used car lot in Miami, Fl and after a few weeks in Sebring, Fl. discovered a couple of rolls of bills tucked between the seats and seat covers. He headed back home to Connecticut.

The jobs available didn't look too appetizing (not to mention the "draft") so he went looking for a recruiter's office to join a service. He wasn't interested in the Army, Marines, or Coastguard.. just the Navy (ships 'n submarines!) and Air Force (airplanes!). The Navy recruiter's office was closed for lunch, so he ended up in the Air Force. Choosing to be an Aircraft Instrument Repairman, he began a 20 year career beginning with the F-105D fighter-bomber nicknamed the "Thud" (that was the sound it made when it hit the ground) and also known as the "triple-threat".... It would bomb, strafe, *and* fall on you! The last aircraft he worked on was the F-16 Fighting Falcon as an Instrument and Flight Controls Repairman, with his last years as an instructor to new troops and foreign troops teaching the repair of multiple systems on and pertaining to the F-16.

During his career he had the privilege to visit, tour, and/or work in Okinawa, Taiwan, Thailand, Japan, South Korea, Philippines, Great Britain, Spain, and Germany (Loved it all and I'd do it again in a heartbeat!).

After "retiring" at the rank of Master Sergeant from the Air Force, he spent four years as a lead technician in a major school district's computer repair facility, left that to go to Arizona to start his own business (failed!). Worked as a cook at Pete's Fish and Chips in Phoenix and ended up driving Class 8 (BIG) trucks with Swift Transportation for 4 years. He then became a licensed financial adviser (couldn't sell a gallon of water to a dehydrated soul comin' off the Gobi Desert, let alone insurance and mutual funds!), and is currently working as a contract calibration technician for the Air Force (and that's just the major stuff!).

His interests are sailing, photography, celestial navigation, Sharia (Look it up, people!) and its infiltration into American society (God help us, because our Government won't!), and, ssshhhhhhh..... guns and "prepping".

Photo 1   Summer of '81. That's my son, Frank, on the step ladder. This was an 8' radius dome.

Photo 2

PHOTO 3

PHOTO 4

Photo 5

PHOTO 7

PHOTO 8a

PHOTO 8B

PHOTO 9

PHOTO 10

PHOTO 11

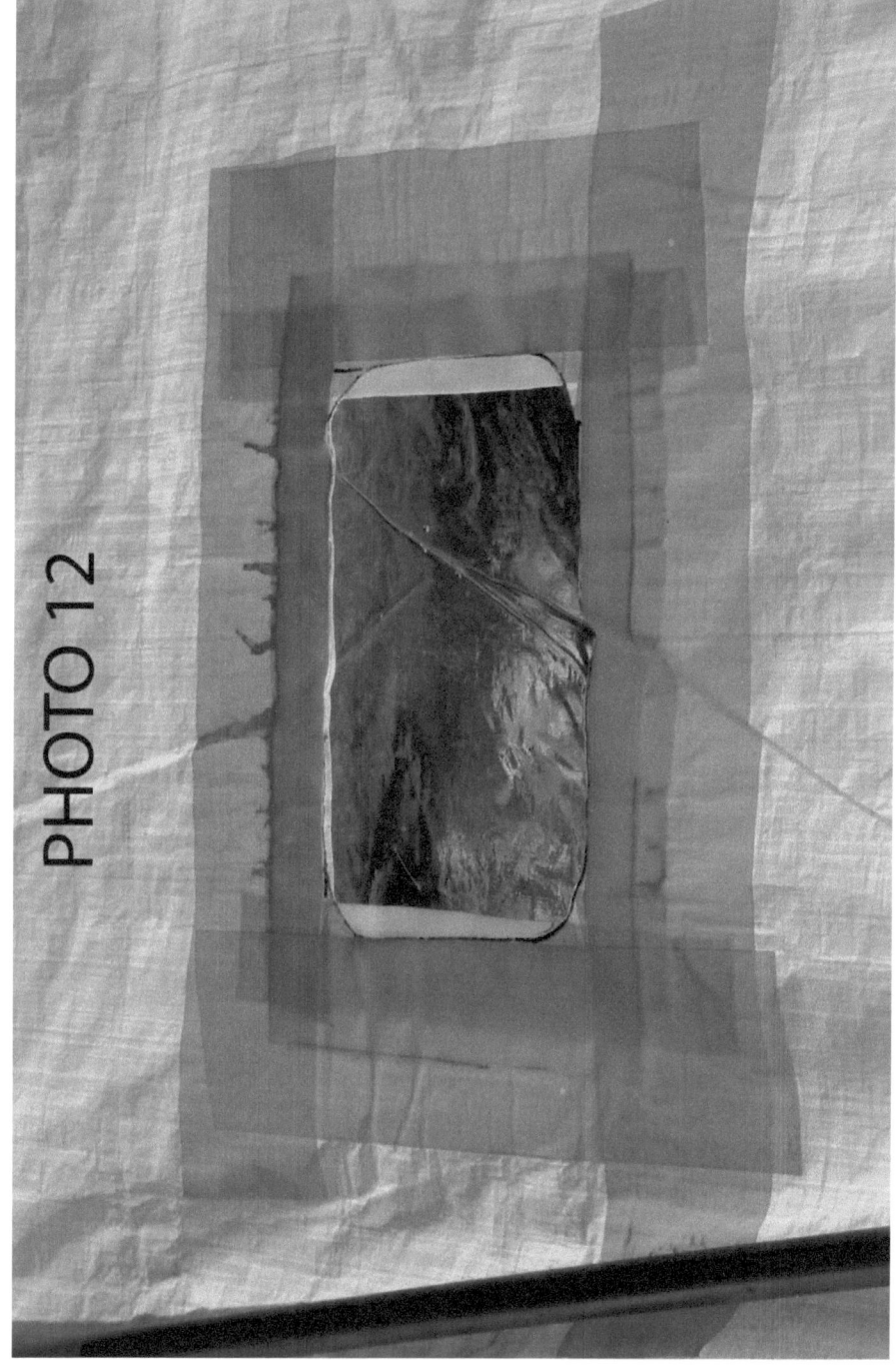

PHOTO 12

PHOTO 14

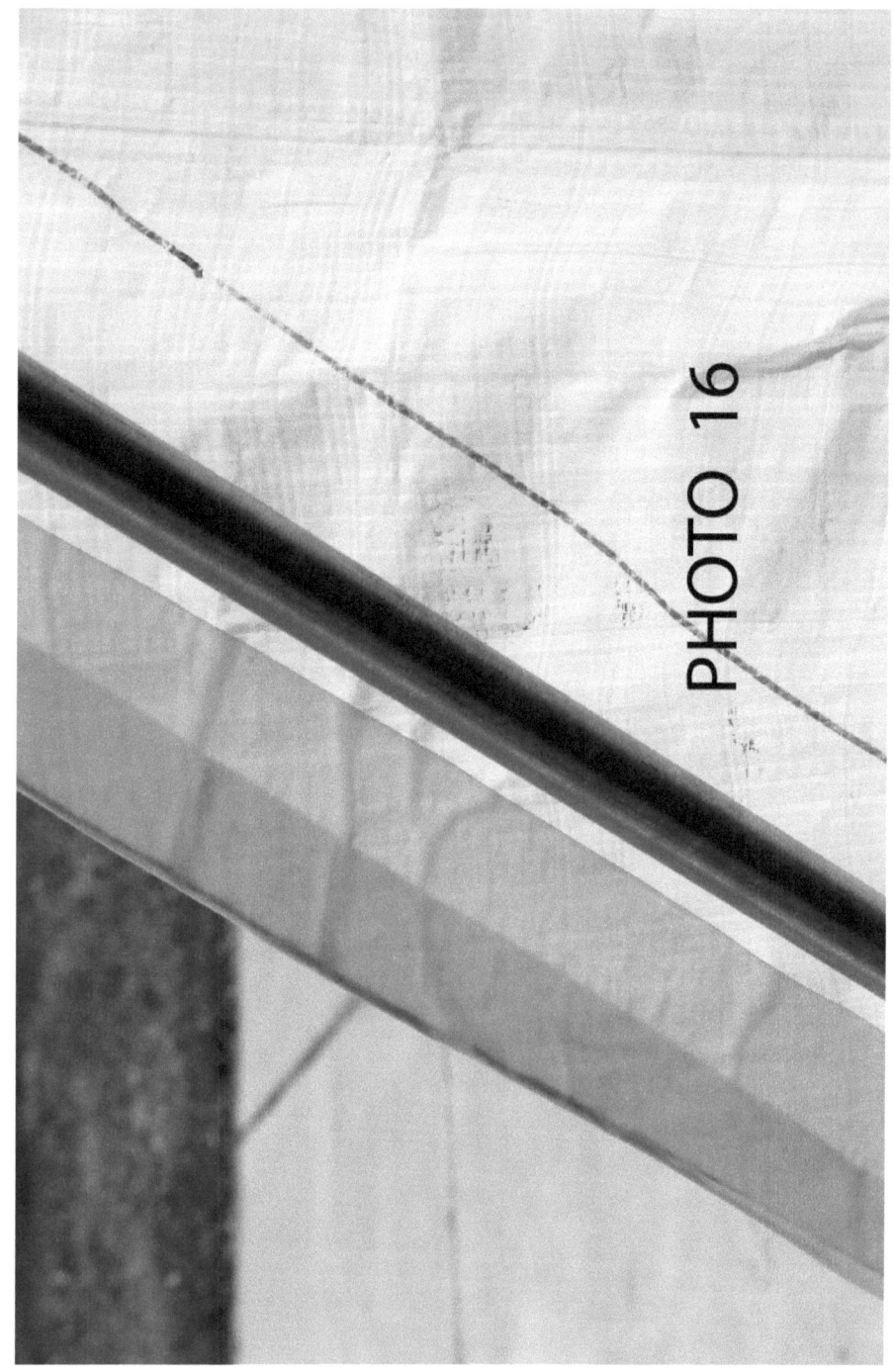

PHOTO 17

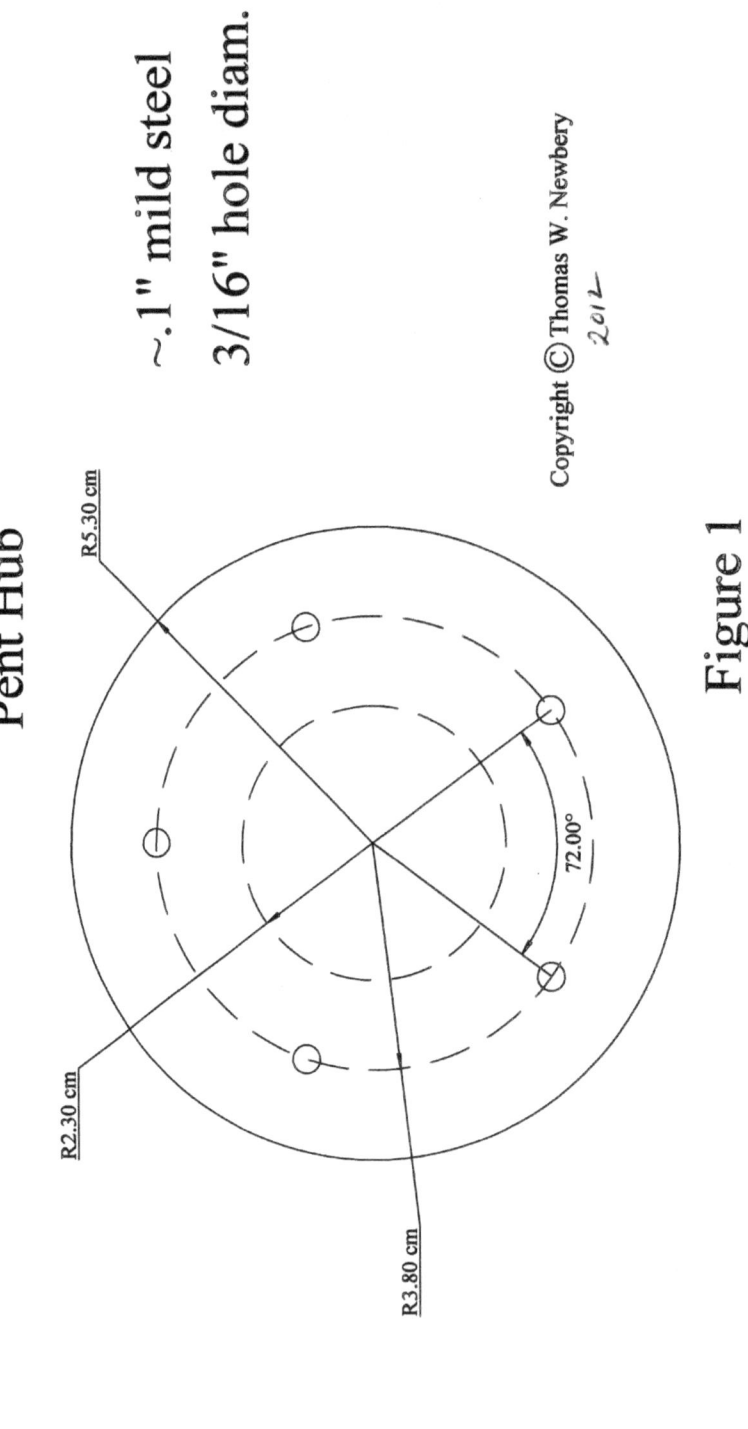

Figure 1

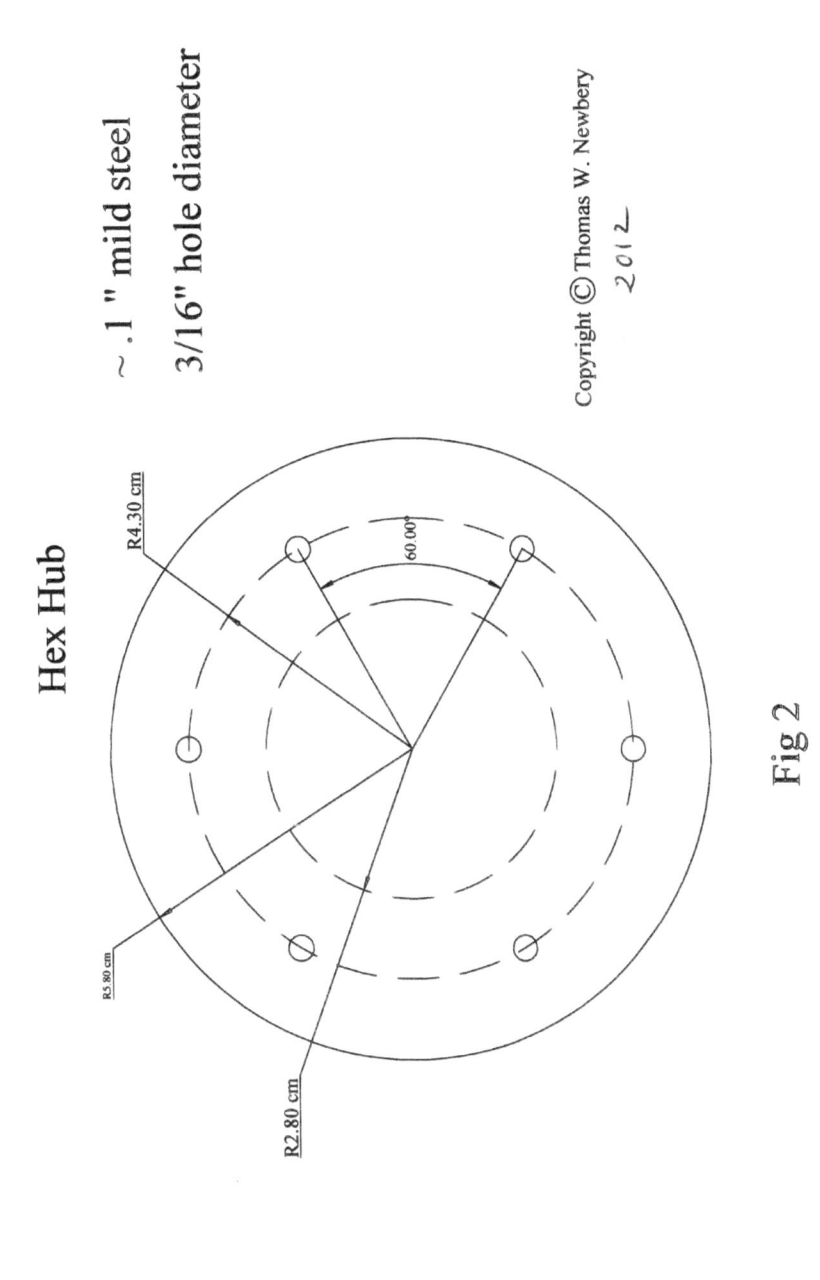

Fig 2

Hex Hub

~ .1 " mild steel

3/16" hole diameter

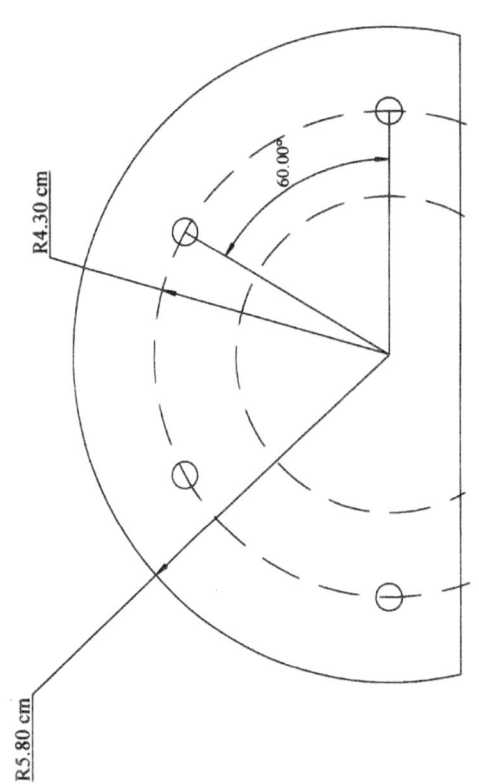

Fig 3

Copyright © Thomas W. Newbery
2012

Short Strut Measurements

OAL (C to C) + 0.5465 × 2.1336 m = 1.1660 m

Chord = 1.1660 m

Raw (cut strut) length = 1.1660 m - .028 m - .023 m
= 1.1660 m - .051 m = 1.115 m

Center to Center Distance Between holes = 1.166 m - .015 m - .028 m - .015 m - .023 m
= 1.166 m - .081 m
= 1.085 m

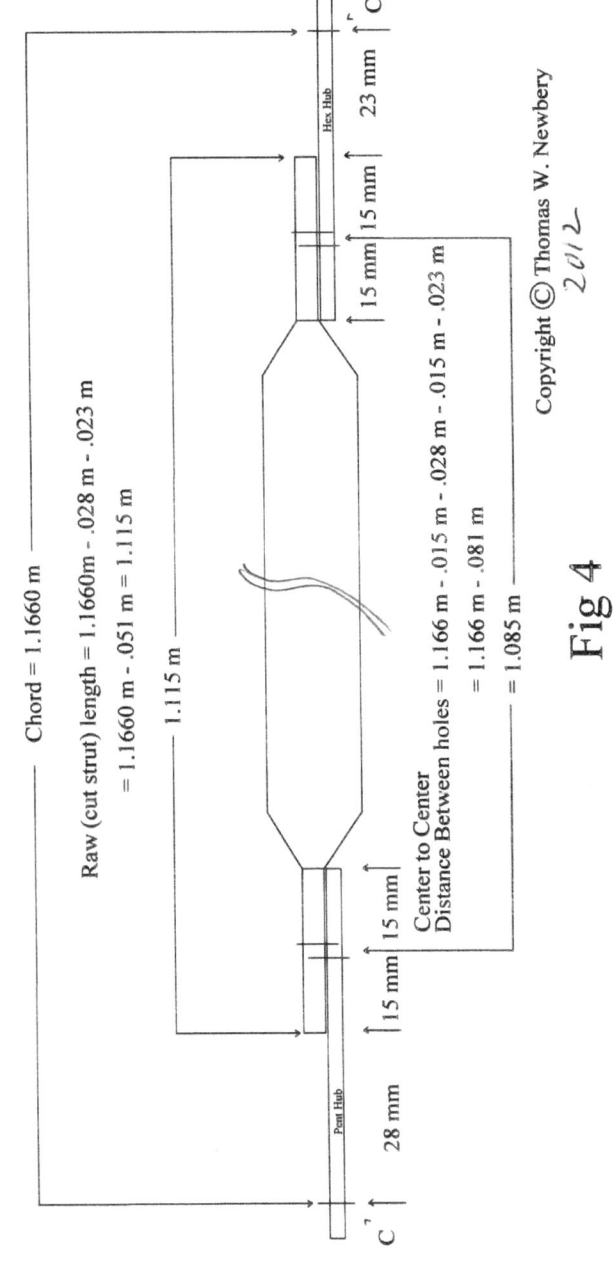

Fig 4

Copyright © Thomas W. Newbery 2012

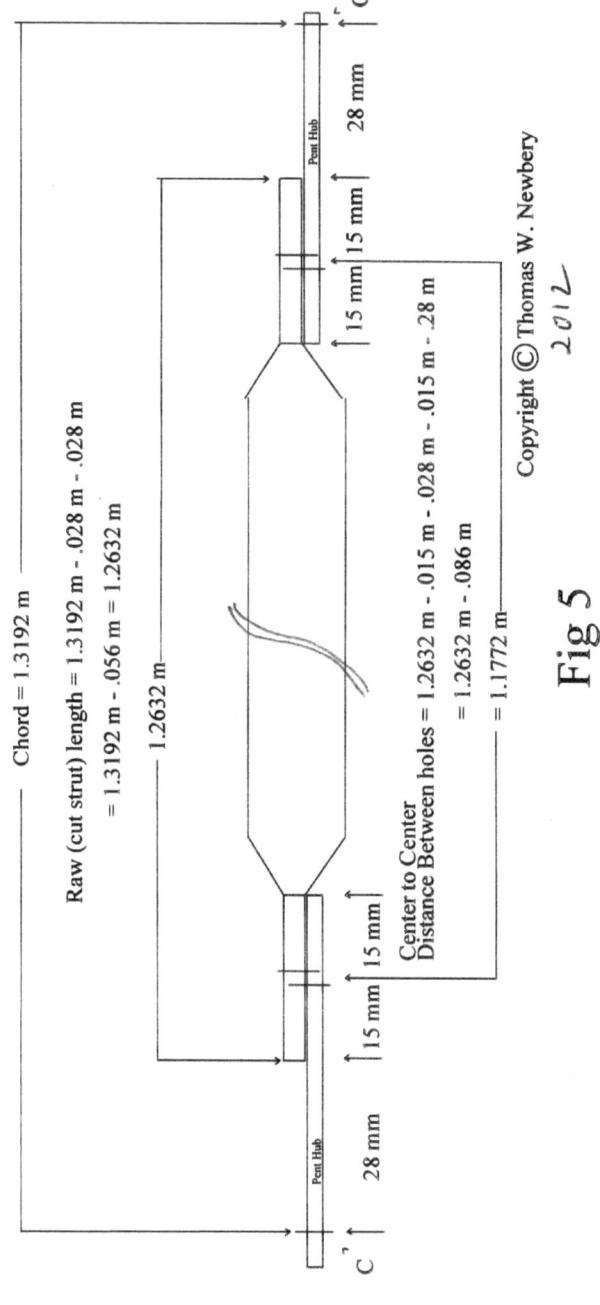

Fig 5

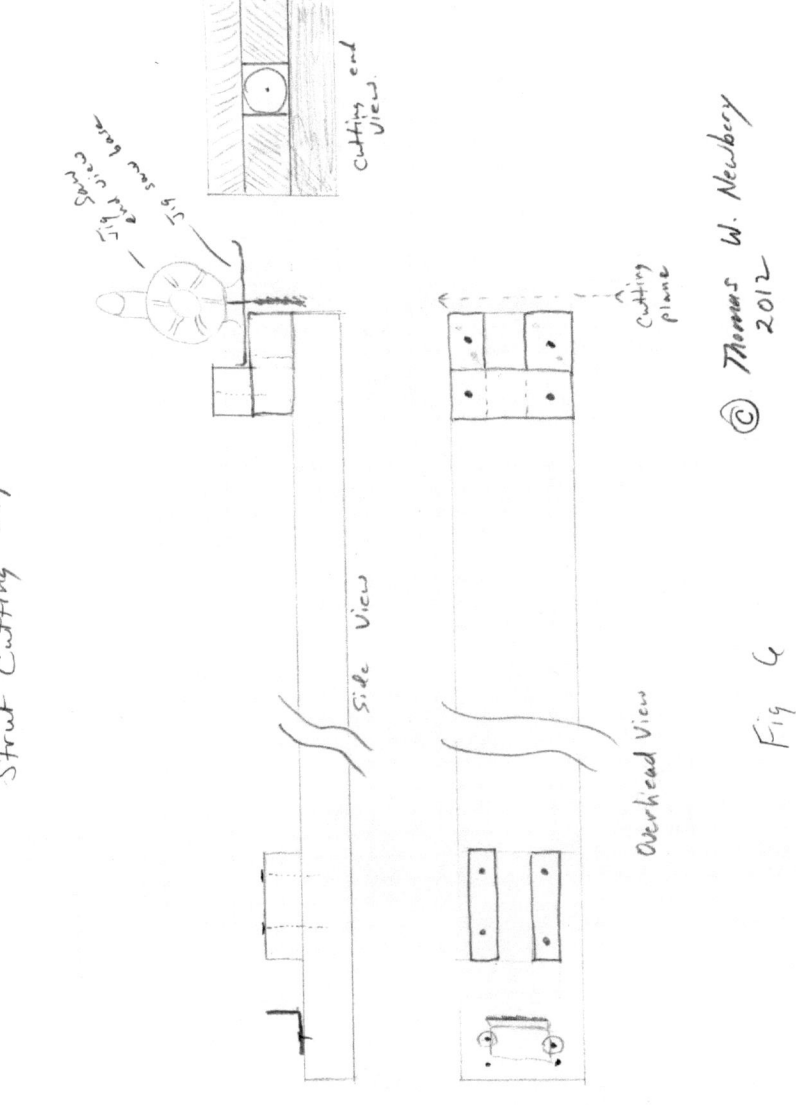

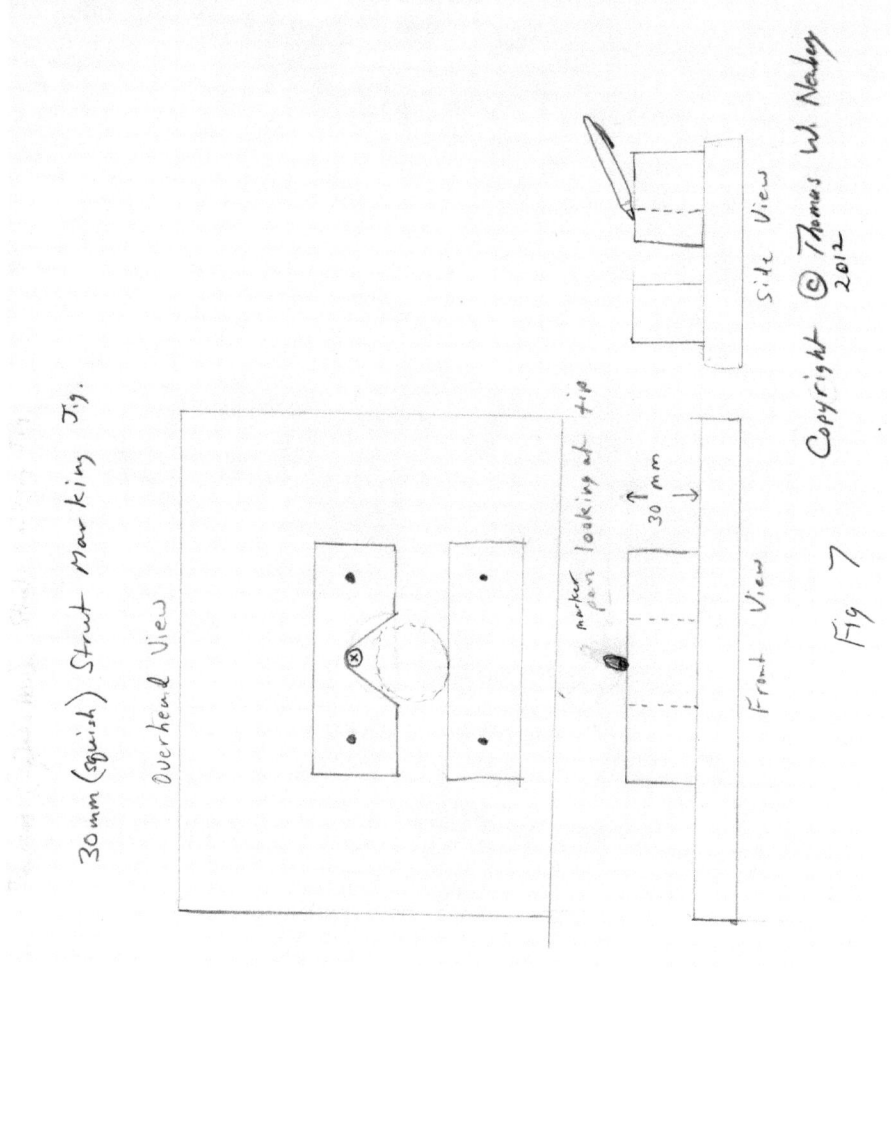

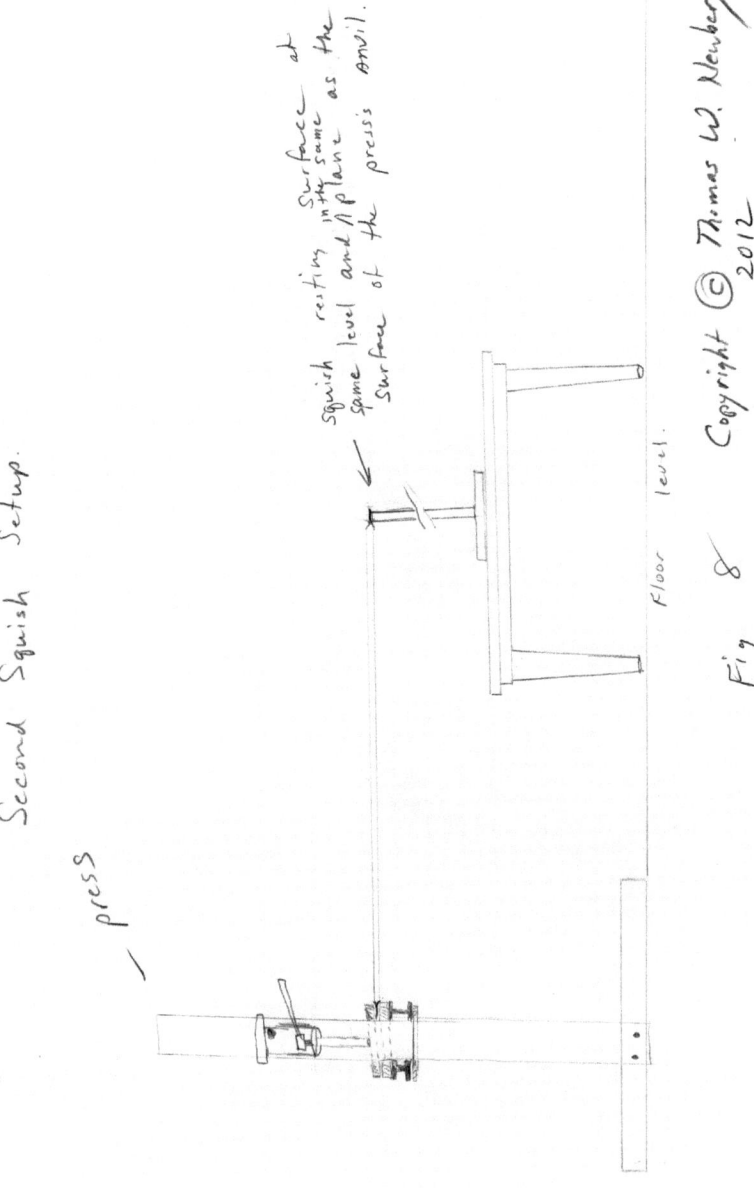

Second Squish Setup.

press

Squish resting Surface at same level and in the same plane as the Surface of the press Anvil.

Floor level.

Fig 8   Copyright © Thomas W. Newberg 2012

First Squish Hole Jig.

⊗ Drill hole well/center of bit.

Affix Base to drill press platform in any convenient manner - C-clamps, bolts, screws - whatever works

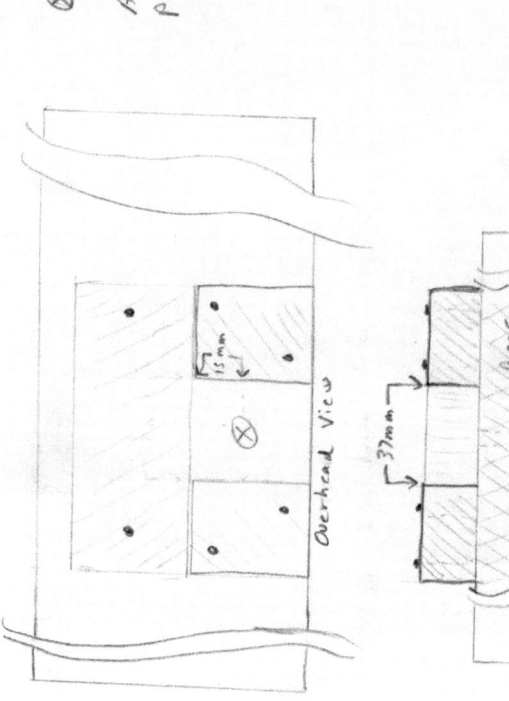

Overhead View

Front View

Base

15mm

37mm

Fig 9. Copyright © Thomas W. Newbery 2012

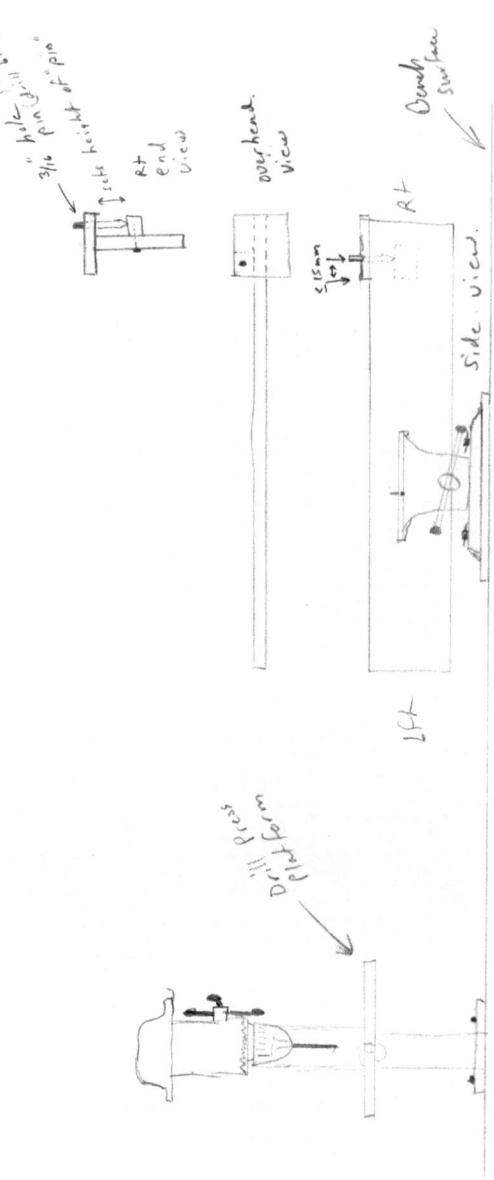

Strut Hole Spacing Jig Using Drill Press Available

Fig 16 Copyright © Thomas W. Newberg 2012

Alternate Hole-to-Hole Jig, no drill press available

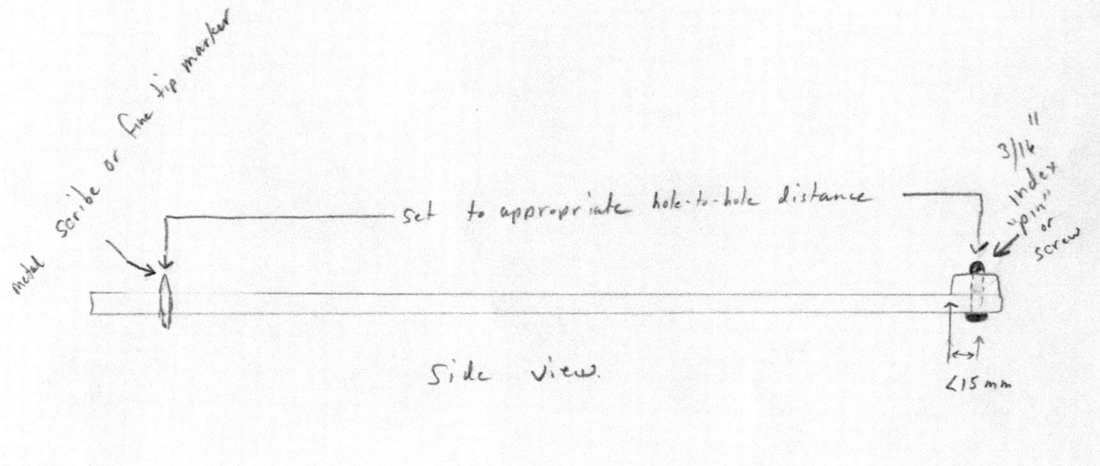

make scribe or fine tip marker

Set to appropriate hole-to-hole distance

Side View

3/4"
1/4" apex
1/8"
screw

<15 mm

Fig 11

Copyright © Thomas W. Newbery
2012

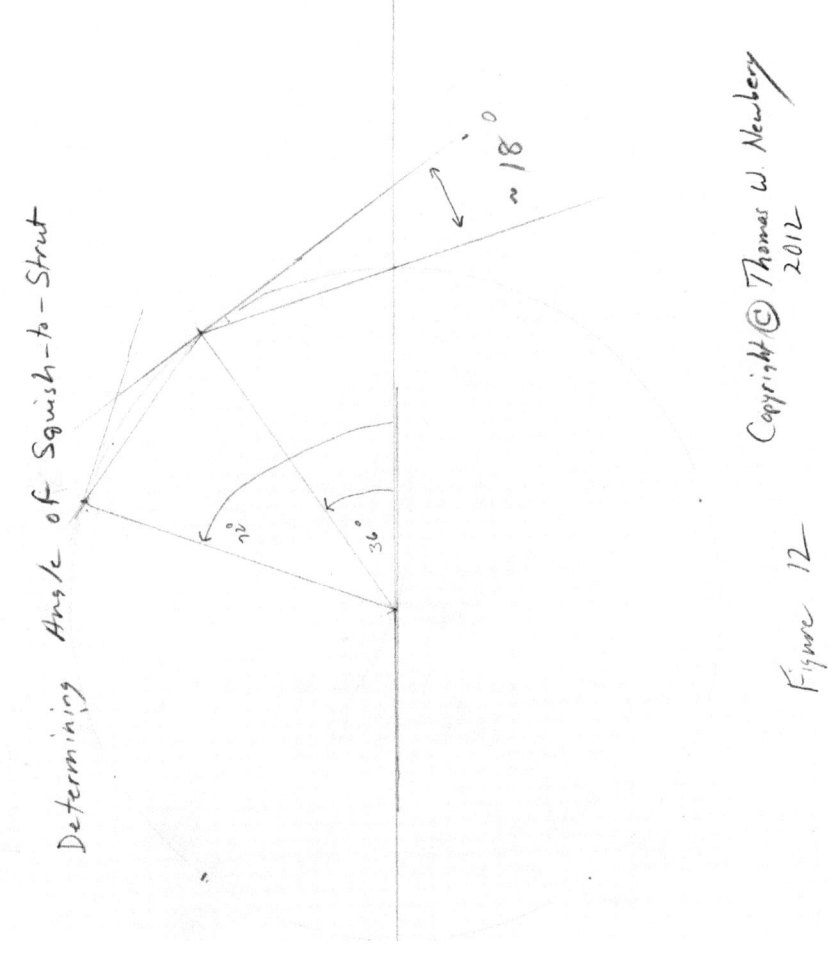

Figure 12 — Determining Angle of Squish-to-Strut

Copyright © Thomas W. Newbery 2012

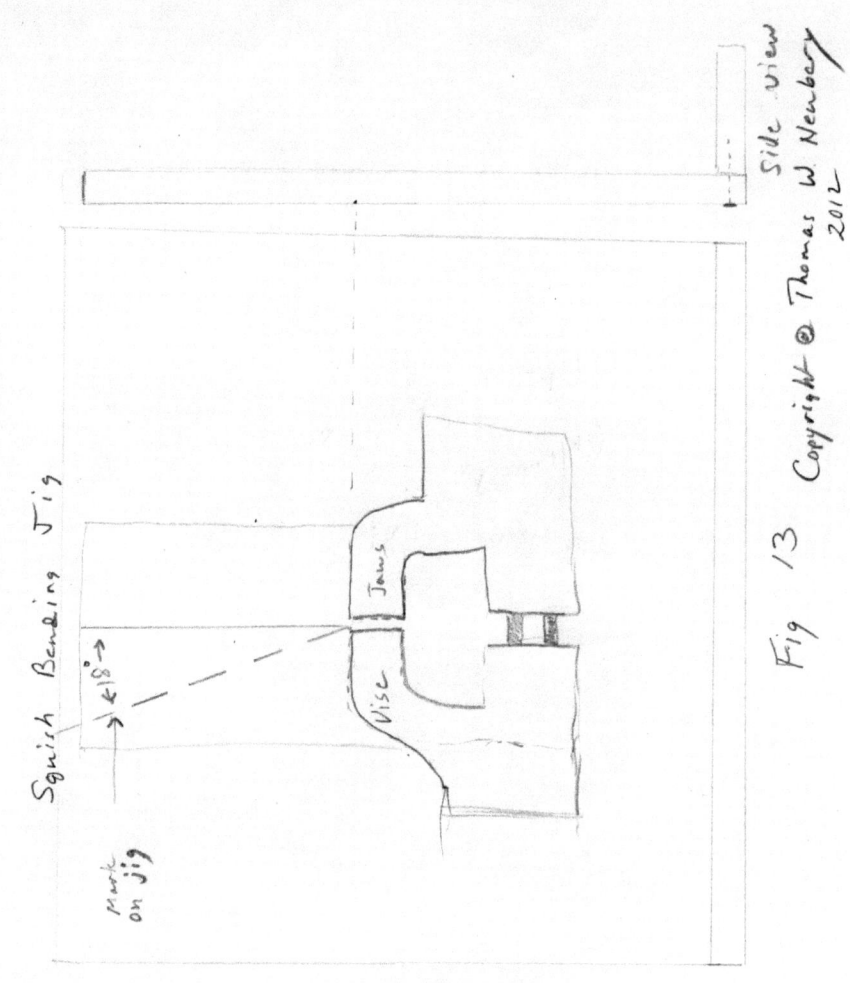

Fig 13 Copyright © Thomas W. Newbery 2012

Wing Nut Tool.

End View

0° Side View

90° Side View

Fig 14 Copyright © Thomas W Newberg 2012

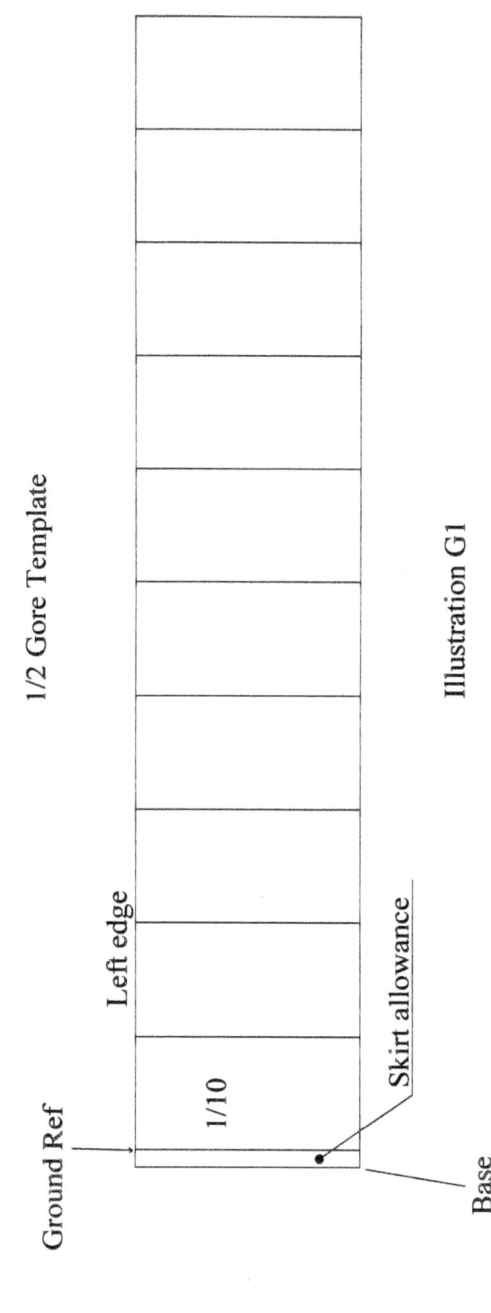

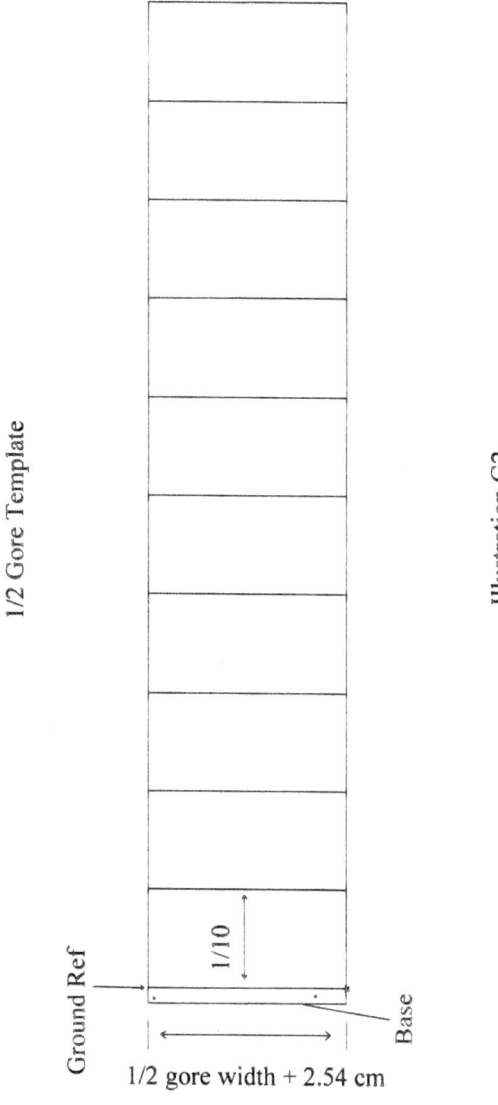

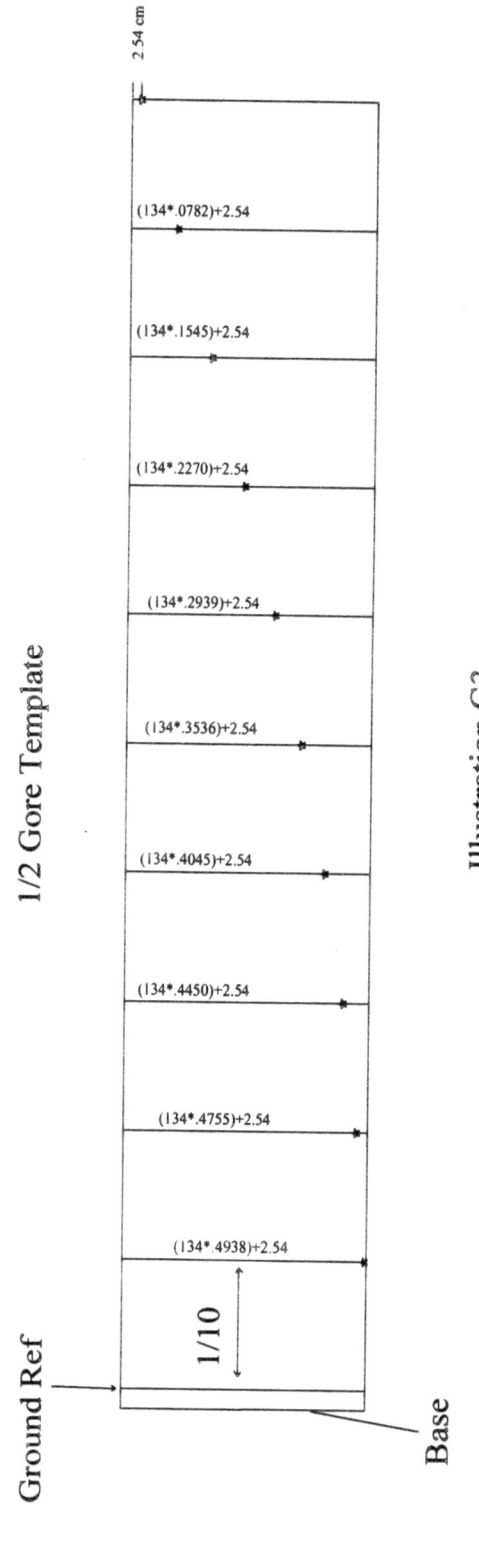

Illustration G3

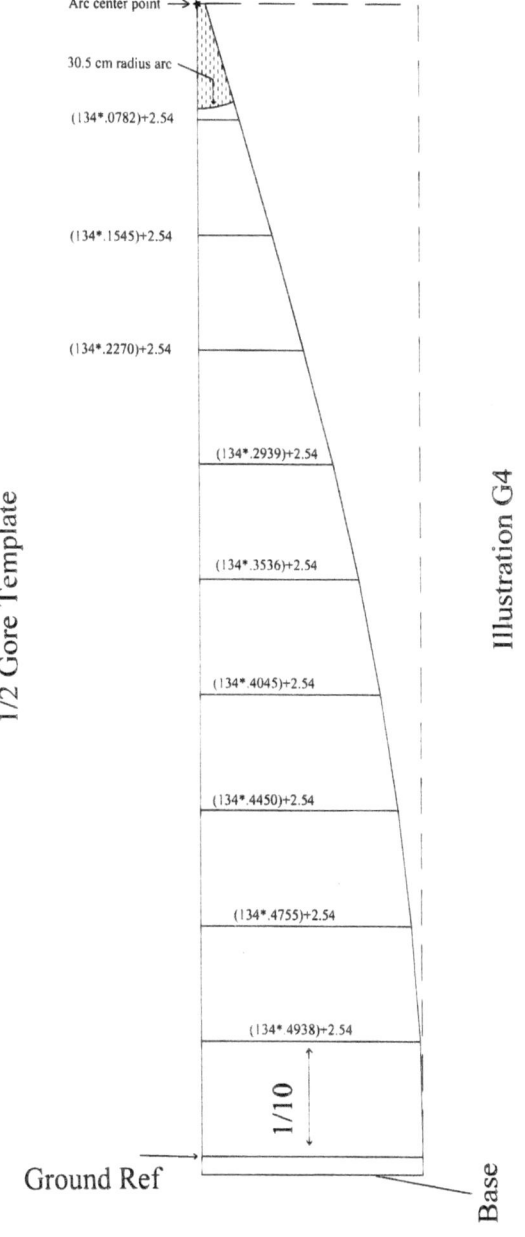

Illustration G4

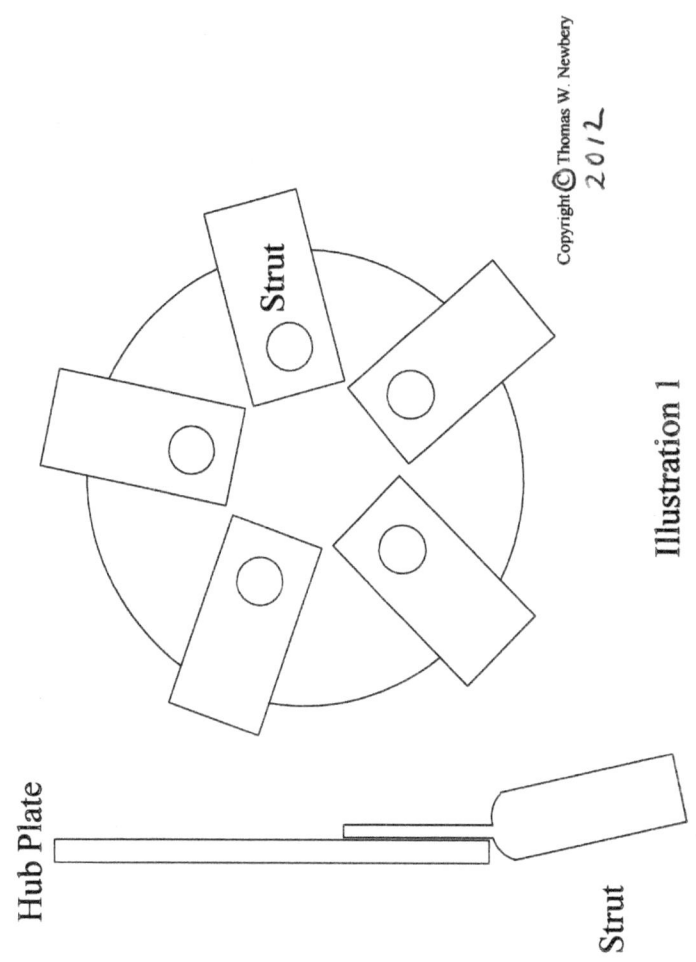

Illustration 1

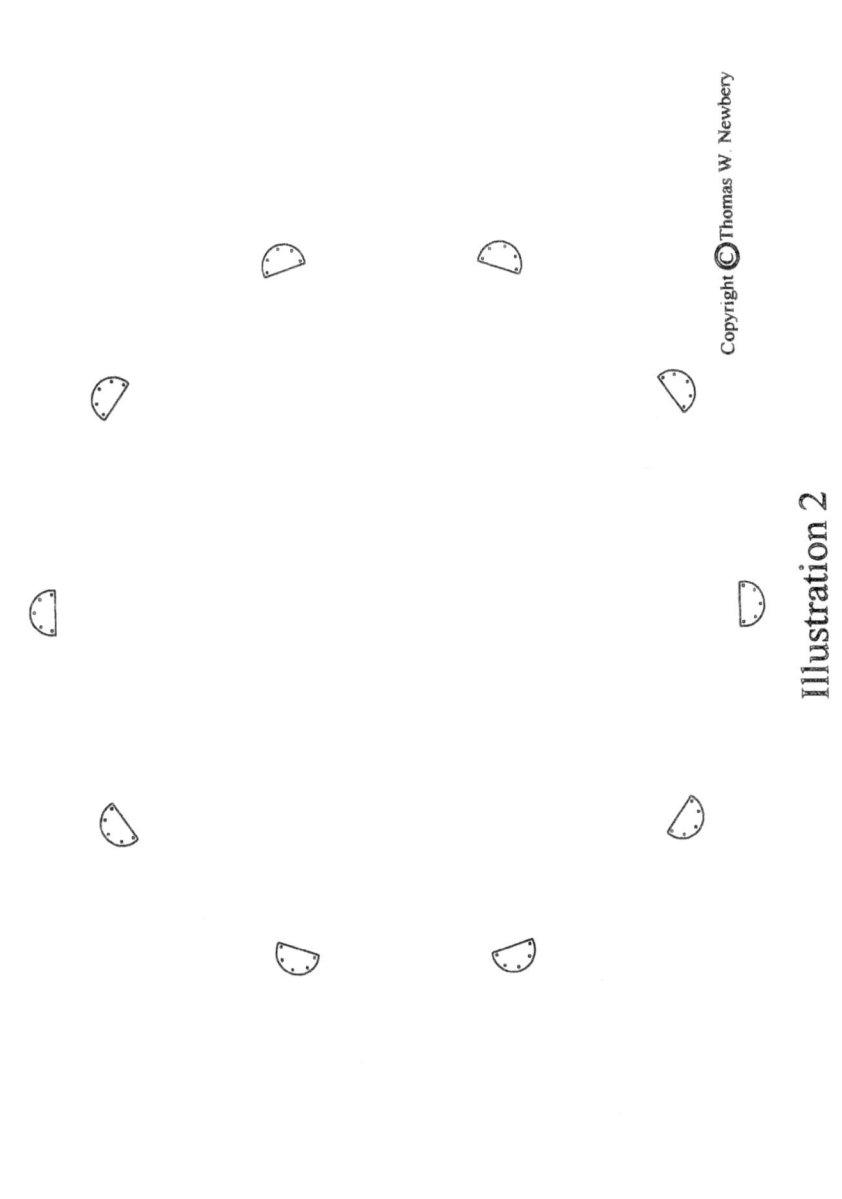

Illustration 2

Copyright © Thomas W. Newbery

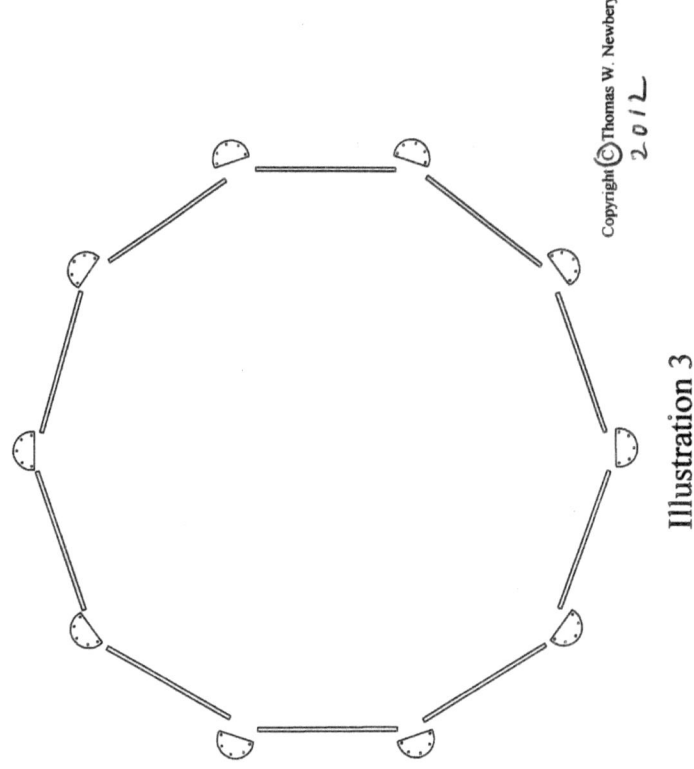

Illustration 3

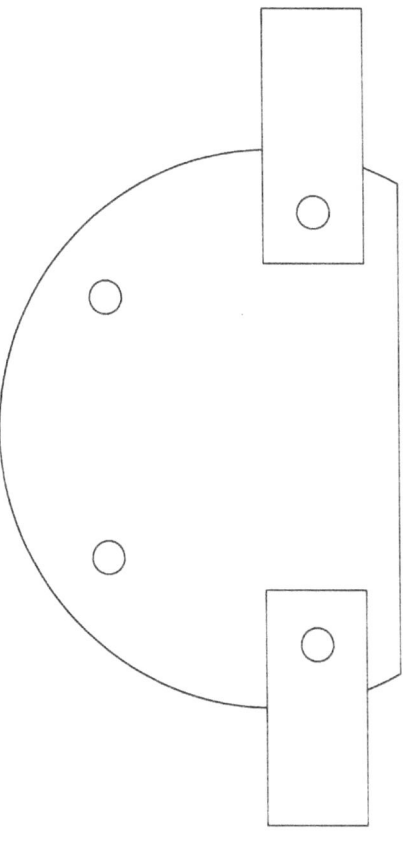

Viewed from inside at ground level, inside

Illustration 4

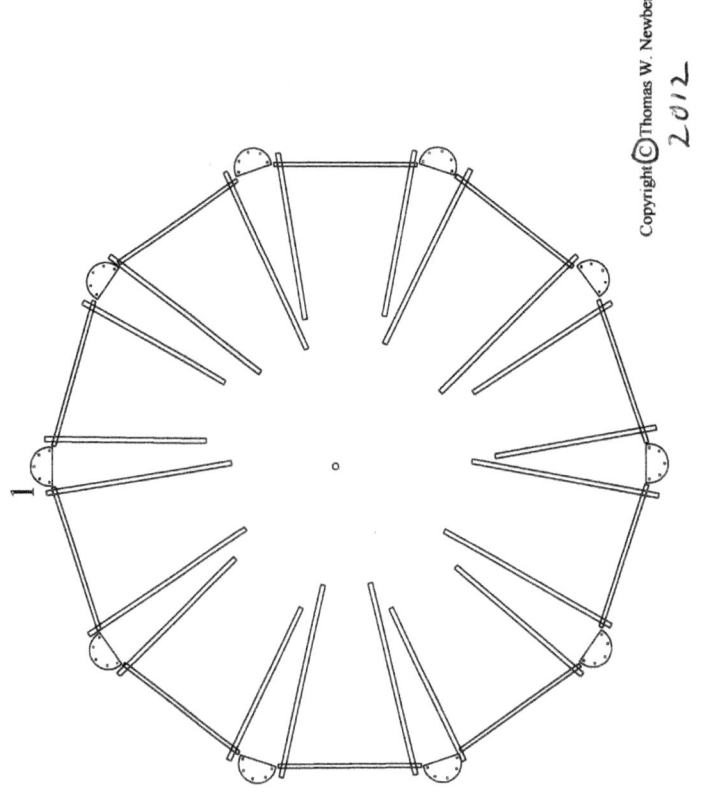

Illustration 5

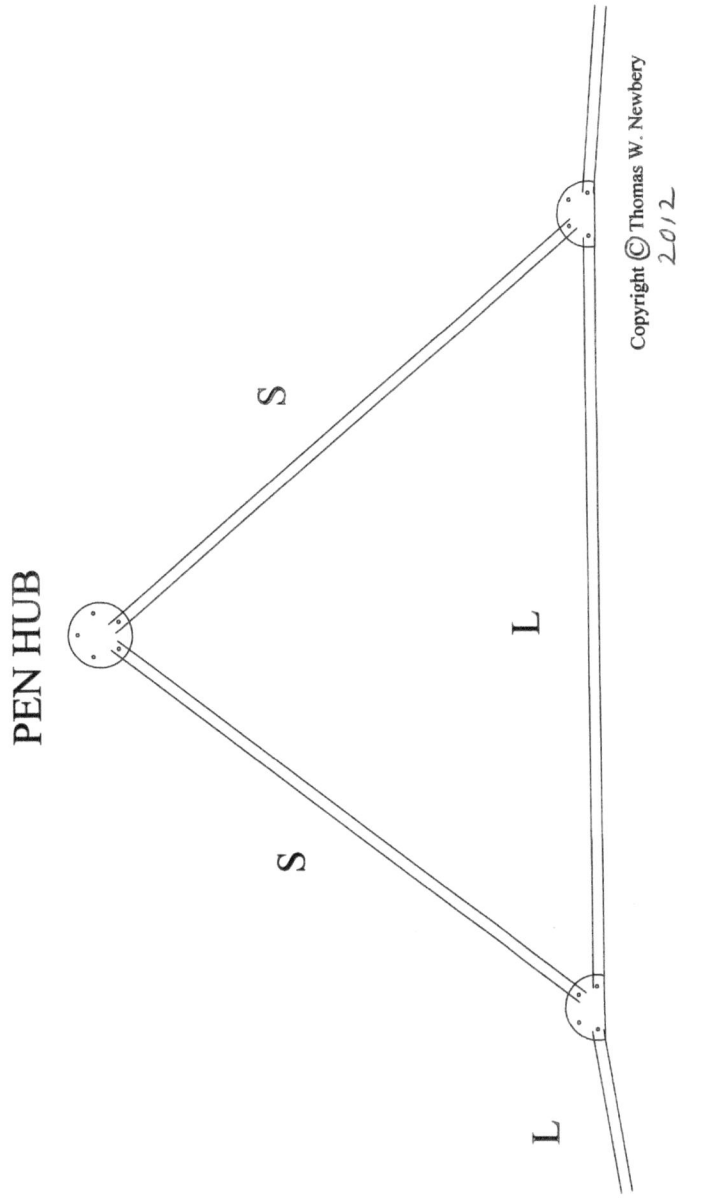

Illustration 6

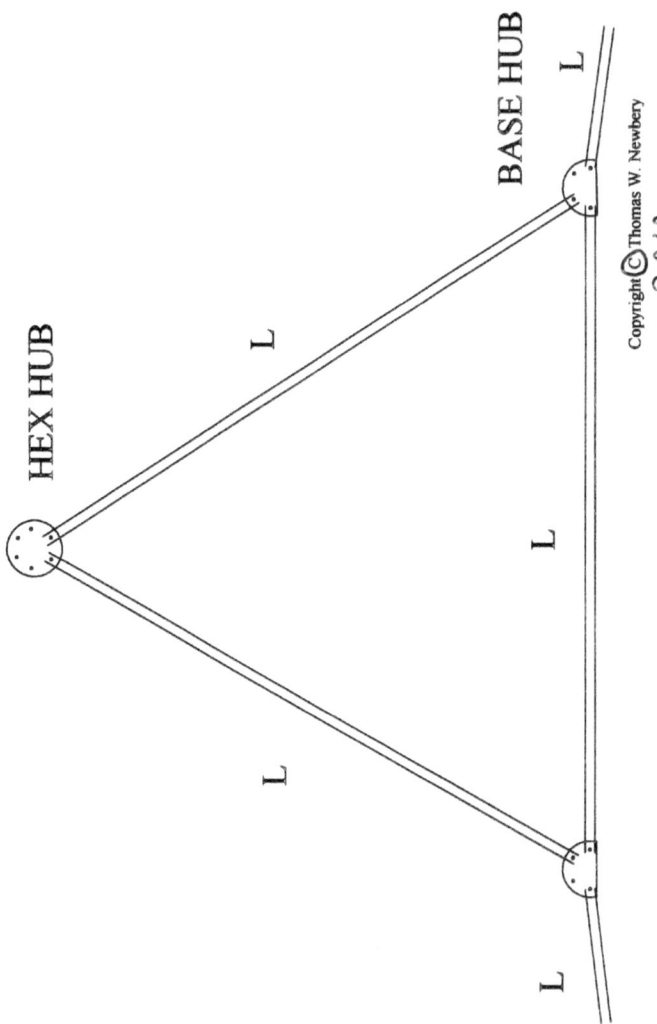

Illustration 7

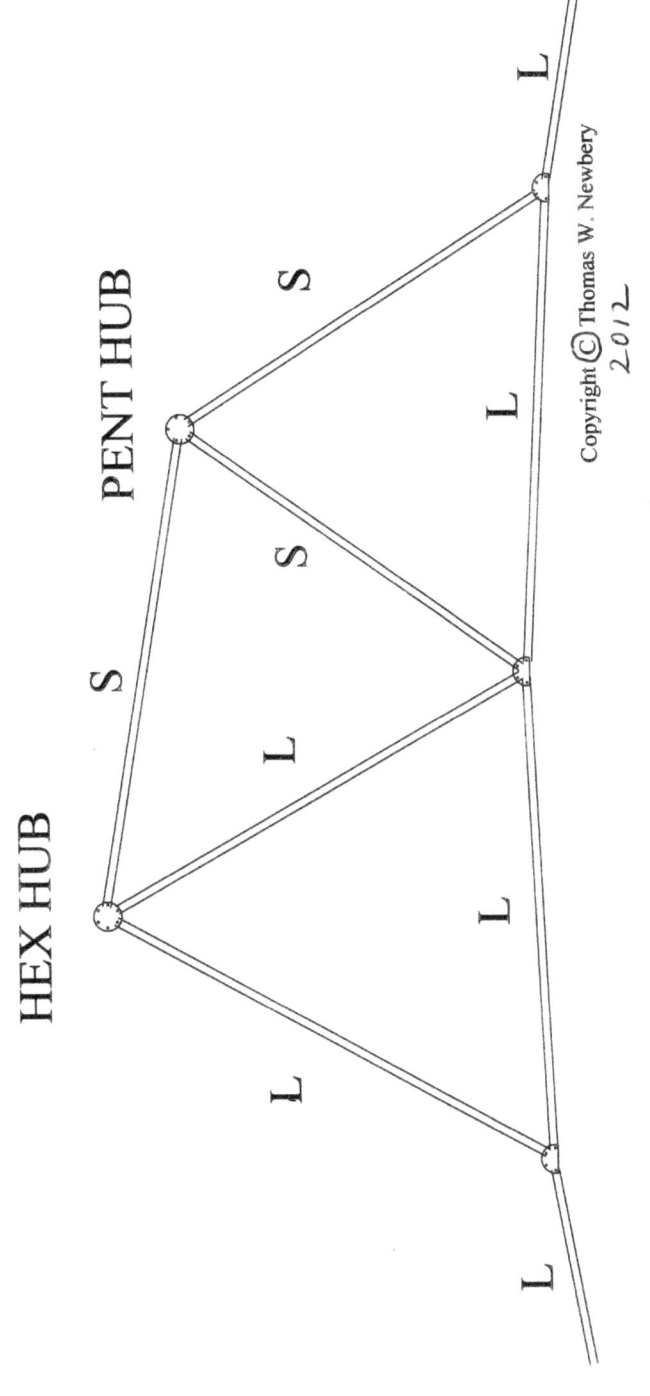

Illustration 8

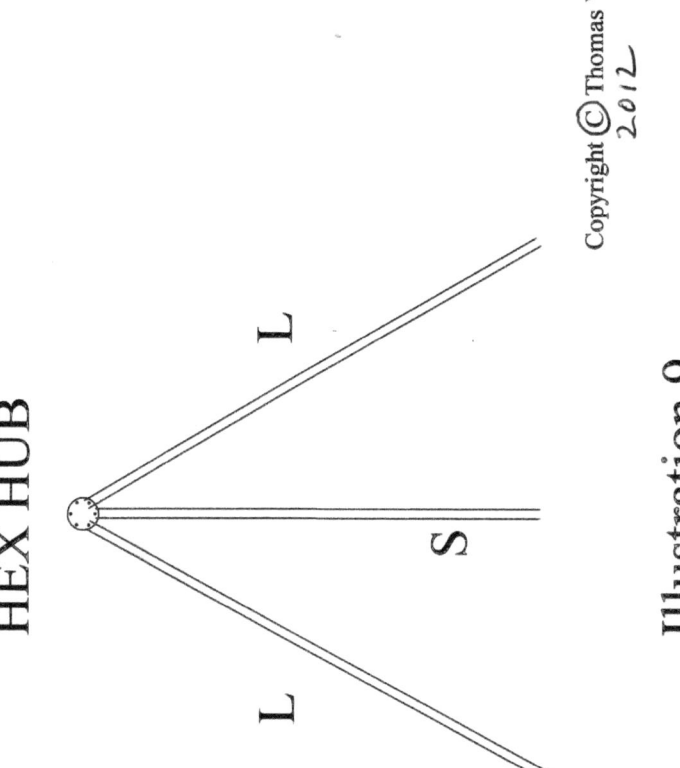

Illustration 9

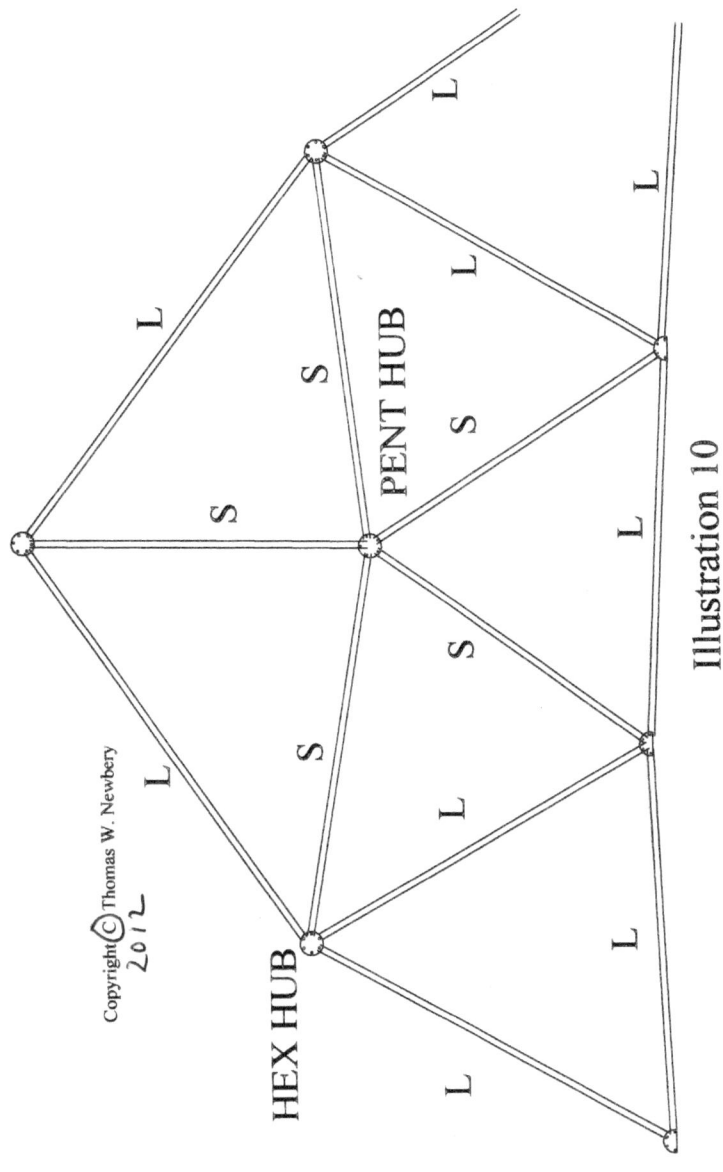

Illustration 10

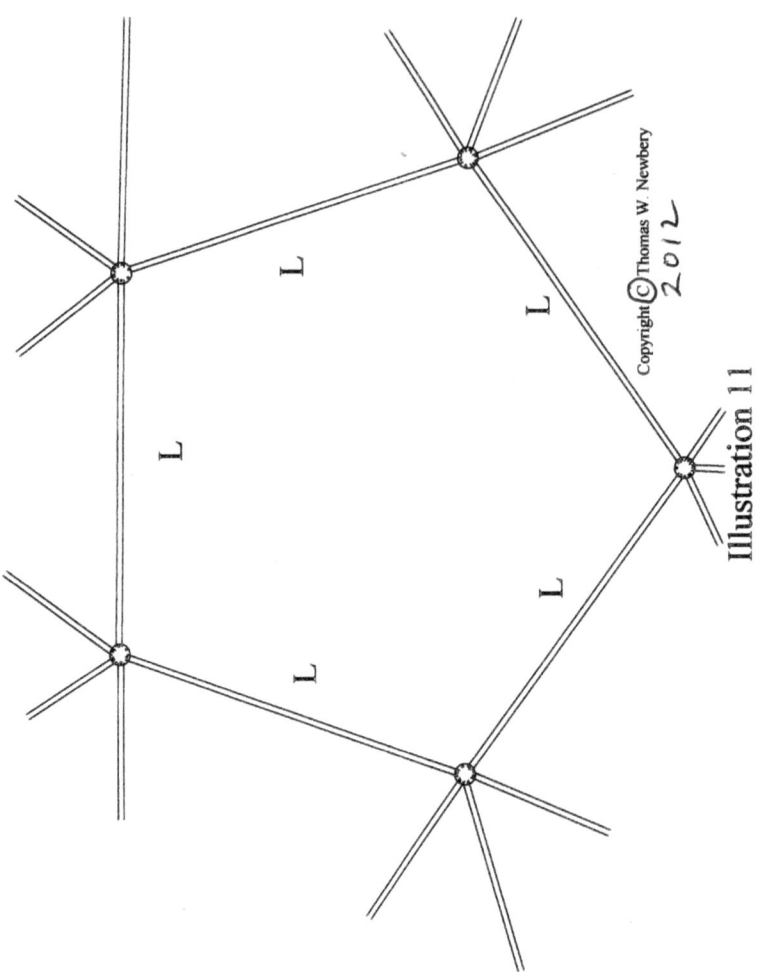

Illustration 11

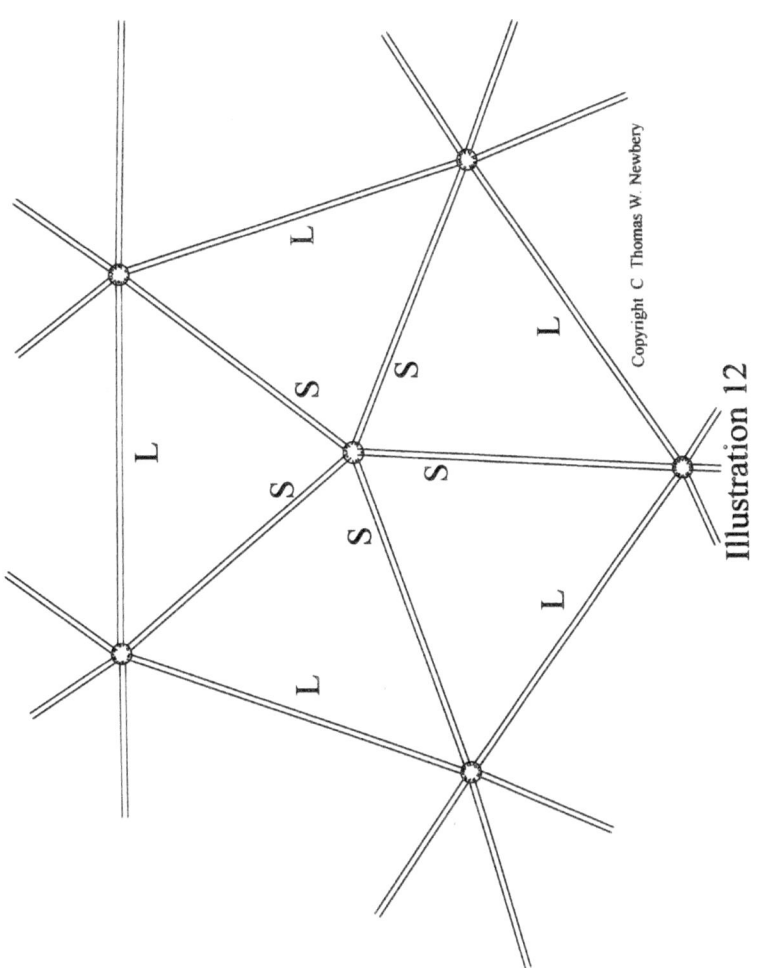

Illustration 12

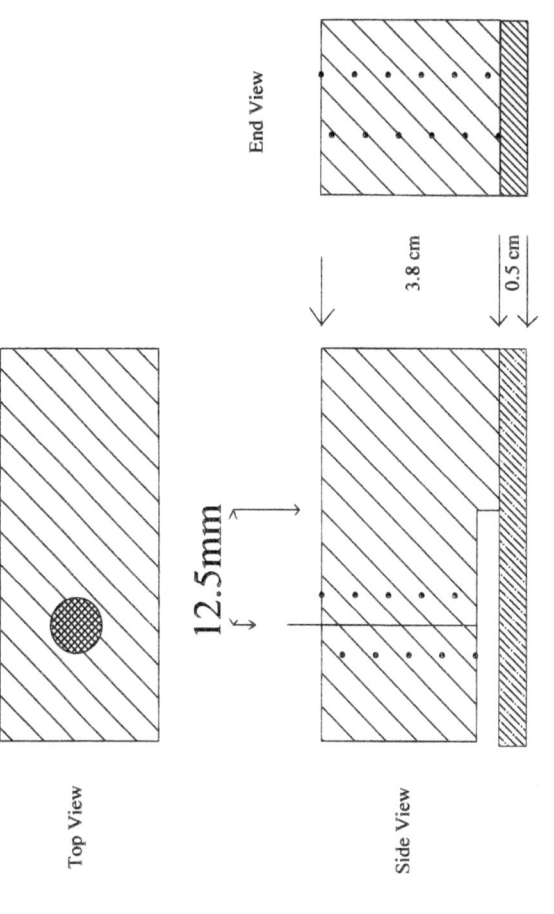

Illustration 13

www.ingramcontent.com/pod-product-compliance
Lightning Source LLC
Chambersburg PA
CBHW050726180526
45159CB00003B/1146